Compact Textbooks in Mathematics

For further volumes:
http://www.springer.com/series/11225

Birkhäuser

Compact Textbooks in Mathematics

This textbook series presents concise introductions to current topics in mathematics and mainly addresses advanced undergraduates and master students. The concept is to offer small books covering subject matter equivalent to 2- or 3-hour lectures or seminars which are also suitable for self-study. The books provide students and teachers with new perspectives and novel approaches. They feature examples and exercises to illustrate key concepts and applications of the theoretical contents. The series also includes textbooks specifically speaking to the needs of students from other disciplines such as physics, computer science, engineering, life sciences, finance.

Hansjoerg Albrecher • Andreas Binder
Volkmar Lautscham • Philipp Mayer

Introduction to Quantitative Methods for Financial Markets

Birkhäuser

Hansjoerg Albrecher
Volkmar Lautscham
Department of Actuarial Science
University of Lausanne
Lausanne
Switzerland

Andreas Binder
Kompetenzzentrum Industriemathematik
Mathconsult GmbH
Linz
Austria

Philipp Mayer
Department of Mathematics
TU Graz
Graz
Austria

Revised and updated translation from the German language edition: Einführung in die Finanzmathematik by Hansjörg Albrecher, Andreas Binder, and Philipp Mayer, © Birkhäuser Verlag, Switzerland 2009. All rights reserved

ISBN 978-3-0348-0518-6 ISBN 978-3-0348-0519-3 (eBook)
DOI 10.1007/978-3-0348-0519-3
Springer Basel Heidelberg New York Dordrecht London

Library of Congress Control Number: 2013940190

2010 Mathematical Subject Classification: 91-01 (91G10 91G20 91G80)

Cover design: deblik, Berlin

Printed on acid-free paper

Springer Basel is part of Springer Science+Business Media (www.springer.com)

Preface

This book is an introductory text to mathematical finance, with particular attention to linking theoretical concepts with methods used in financial practice. It succeeds a German language edition, Albrecher, Binder, Mayer (2009): Einführung in die Finanzmathematik. Readers of the German edition will find the structures and presentations of the two books similar, yet parts of the contents of the original version have been reworked and brought up-to-date. Today's financial world is fast-paced, and it is especially during financial downturns, as the one initiated by the 2007/08 Credit Crisis, that practitioners critically review and revise traditionally employed methods and models.

The aim of this text is to equip the readers with a comprehensive set of mathematical tools to structure and solve modern financial problems, but also to increase their awareness of practical issues, for instance around products that trade in the financial markets. Hence, the scope of the discussion spans from the mathematical modeling of financial problems to the algorithmic implementation of solutions. Critical aspects and practical challenges are illustrated by a large number of exercises and case studies.

The text is structured in such a way that it can readily be used for an introductory course in mathematical finance at the undergraduate or early graduate level. While some chapters contain a good amount of mathematical detail, we tried to ensure that the text is accessible throughout, not only to students of mathematical disciplines, but also to students of other quantitative fields, such as business studies, finance or economics. In particular, we have organized the text so that it would also be suitable for self-study, for example by practitioners looking to deepen their knowledge of the algorithms and models that they see regularly applied in practice.

The contents of this book are grouped in 15 modules which are to a large degree independent of each other. Therefore, a 15-week course could cover the book on a one-module-per-week basis. Alternatively, the instructor might wish to elaborate further on certain aspects, while excluding selected modules without majorly impairing the accessibility of the remaining ones. Conversely, single modules can be used separately as compact introductions to the respective topic in courses with a scope different from general mathematical finance.

Due to its compact form, we hope that students will find this book a valuable first toolbox when pursuing a career in the financial industry. However, it is obvious that there exists a wide range of other methods and tools that cannot be covered in the present concise format and some readers might feel the need to study some aspects in more detail. To facilitate this, each module closes with a list of references for further reading of theoretical and practical focus. The reader is furthermore encouraged to check his/her understanding of the covered material by solving exercises as listed at the end of each module, and to implement algorithms to gain experience in implementing solutions. Some of the exercises further develop presented techniques and could also be included in the course by the instructor.

In terms of prior knowledge, the reader of this book will find some understanding of basic probability theory and calculus helpful. However, we have tried to limit any prerequisites as much as possible. To link the concepts to practical applications, we aimed at making the reader comfortable with a certain scope of technical language and market terms. Technical terms are printed in *italics* when used for the first time, whilst terms introducing a new subsection are printed in **bold**. To improve the text's readability, additional information is provided in footnotes in which one will also find biographic comments on some persons who have greatly contributed to developing the field of mathematical finance.

Several algorithmic aspects are illustrated through examples implemented in Mathematica and in the software package UnRisk PRICING ENGINE (in the following: UnRisk). UnRisk (www.unrisk.com) is a commercial software package that has been developed by MathConsult GmbH since 1999 to provide tools for the pricing of structured and derivative products. The package is offered to students free of charge for a limited period post purchase of this book. UnRisk runs on Windows engines and requires Mathematica as a platform.

We hope that you will enjoy assembling your first toolbox in mathematical finance by working through this book and look forward to receiving any comments you might have at quantmeth.comments@gmail.com.

Lausanne, Linz and Brussels,
April 2013

Hansjörg Albrecher, Andreas Binder,
Volkmar Lautscham and Philipp Mayer

Contents

Interest, Coupons and Yields

1

Each of us has experience with paying or receiving *interest*. If you wish to purchase goods today despite having insufficient funds, you can, for example, borrow money from a bank. Your desired purchases could include a house, a car or consumption goods, and the borrowing could be in the form of a current account overdraft or a *term loan*. You take the position of a *borrower*, while the bank acts as *creditor* (or: *lender*) and it will charge you interest on the amount you owe.

On the other hand, when you have accumulated savings that you wish to spend only in the future, you can lend the money to banks (in the form of deposits), governments (government bonds), or corporations (corporate bonds), which will pay you interest on the funds provided.

In the retail saving-lending market, banks take the position of *financial intermediaries*. Financial intermediaries have many functions, including *size transformation* (many small deposits can be accumulated to provide one large loan to e.g. a corporate) and *term transformation* (small short-term deposits can be transformed into a longer-term loan).

1.1 Time Value of Money

An investor providing funds to a borrower will expect to receive a financial return, and if the money is provided as debt, the return will be in the form of *interest payments*. How much interest is paid will depend, among other factors, on the borrowed amount, the time until repayment (or: *maturity*) and on the likelihood of the borrower making payments in the future as agreed in the loan contract.

Assuming liquid financial markets, unrestricted mobility of capital and complete information for all market participants would imply that borrowers of identical credit quality pay the same amount of interest for identical loan structures (including the same starting date, term, borrowed amount, and currency). However, this is not entirely the case in practice. The reasons include that the capital of retail investors is not sufficiently mobile to choose the best investment between all investments

H. Albrecher et al., *Introduction to Quantitative Methods for Financial Markets*,
Compact Textbooks in Mathematics, DOI 10.1007/978-3-0348-0519-3_1,

available, and the fact that certain investments are treated with tax advantages, such as certain pension saving products.

The part of the interest costs in excess of what is charged for otherwise identical but (quasi) risk-free structures, is sometimes referred to as *credit spread*. Debt issues by governments of stable developed economies (e.g. the US, Germany or the UK) are often priced close to risk-free, whereas private borrowers, such as individuals or corporations, might pay significantly higher interest. The risk that the borrower will not make contractual payments in full and on time is called *credit risk* (cf. Chapter 15 – we will neglect credit risk until then).

Suppose the amount $B(t_0)$ is invested at time t_0 (measured in years) for a term of one year. The borrower agrees to pay an interest rate of $R\,\%$ per year (also: *per annum, p.a.*). After a year the borrower will repay $B(t_0) \cdot (1 + R/100)$ under the loan agreement. The balance of the lender's cash account in one year from t_0 (after interest payment and repayment of the borrowed amount) would therefore be

$$B(t_0 + 1) = B(t_0) \cdot \left(1 + \frac{R}{100}\right).$$

1.2 Interest on Debt, Day-Count Conventions

Debt products with a maturity in excess of one year often offer at least annual cash payments. Such products include loans from banks, and *bonds* as their capital market counterparts. Bonds are debt securities that promise the payment of some *principal amount* and regular (e.g. annual) *coupons*[1] (see Section 2.1).

Example

Bond terms of a bond issue by the Government of Austria "2006-2016/2/144A (1st extension)" with security code ISIN AT0000A011T9 (source: Austrian control bank)
Borrower: Republic of Austria
Issue volume: 1.65bn EUR
Issue date: 7 July 2006
Maturity date: 15 September 2016 (10 years 70 days)
Coupon payments: 4 % p.a. on the principal amount, annual coupon
First coupon payment day: 15 September 2006
Day-count convention: ACT/ACT; business-day convention: TARGET

[1] In earlier days bond investors physically held certificates promising the coupon payments and principal repayments. To receive interest payments the investor would exchange coupons against cash on the payment dates. The coupons came in the form of stubs attached to the main bond certificate. Nowadays bond certificates are typically held by trustees and payments are made based on electronic registration systems.

As not many investors would be able to provide the entire amount raised in a corporate or government bond issue, such issues are typically split into many small bonds that can be distributed to a large number of investors. The *principal amount* (or: *nominal, face value*) of such a bond could, for instance, be 1,000 EUR or 10,000 EUR.[2] The market place where investors can buy bonds in a new bond issue is called *primary market*. The splitting of a bond issue into smaller bonds will increase the number of potential buyers, and also ensure liquidity when primary market investors wish to sell on their bonds to other investors in the *secondary market* at a later time prior to maturity.

Note that a capital market investor would not necessarily pay face value (or: *at par*) for a bond initially. If investors see the coupon payment, of e.g. 4% p.a., as too low (high), they will offer less (more) than face value.[3]

The actual coupon payment on a payment date is determined by the nominal interest rate $R\%$ (here: 4% p.a.) times the fraction of a year since the last coupon payment date under a specified *day-count convention*. Denote the day from which interest is accrued as $t_1 = (D1/M1/Y1)$, the date up to which interest is accrued as $t_2 = (D2/M2/Y2)$, and the number of interest bearing days as D_i. When calculating D_i for an interest period $(t_1, t_2]$, the first day is typically excluded and the last day is included, so that no days are double-counted. Widely used day-count conventions include the following.[4]

- '30/360': $D_{30/360} = (D2 - D1) + (M2 - M1) \times 30 + (Y_2 - Y_1) \times 360$ and the coupon payment at t_2 is

$$\text{principal} \times \frac{R}{100} \times D_{30/360}/360.$$

  Note that, in principle, months are equally weighted in the 30/360 method, despite having a different number of days.[5] 30/360 is the typical method used for US government bonds.
- 'Actual/365': days are counted as they occur. $D_{\text{Actual}/365}$ = number of days between t_1 and t_2, so that the coupon payment at t_2 is given by

$$\text{principal} \times \frac{R}{100} \times D_{\text{Actual}/365}/365.$$

---

[2]We will refer to currencies by their three-letter ISO 4217 codes as used in currency trading, for example EUR, GBP, USD, CHF, JPY, SEK.

[3]If a bond with a face value of 100 trades at 100, it is said to *price at par*. If it trades below 100, one would say that it trades at a *discount to face value*, and for prices of above 100 we would say it trades at a *premium to face value*

[4]For further details check, for example, SWX Swiss Exchange [17].

[5]When using a 30/360 method, there are different conventions of counting when e.g. D2 = 31 and D1 = 30.

Note that over a leap year the interest paid is principal $\times R/100 \times 366/365$. In practice you can also find 'Actual/Actual', where the number of days in a leap year is divided by 366 and days in non-leap years are divided by 365, so that the interest paid in 365 and 366-day years is equal.

- 'Actual/360': days are counted the same way as in the previous example, i.e. $D_{\text{Actual}/365} = D_{\text{Actual}/360}$, but coupons are generally higher, at

$$\text{principal} \times \frac{R}{100} \times D_{\text{Actual}/365}/360.$$

This is also called 'French' method and is widely used in the *money markets* (i.e. for maturities not exceeding one year, including USD and EUR markets) and for EUR mortgages.

Further to the government bond example, note that 15 September 2007 was a Saturday and coupon payments are typically only made on business days. How to deal with such a case is agreed upon in the *business-day conventions*. *Modified following* is a popular choice, and defines that coupon payments are carried out on the day if it is a business day, or otherwise on the first business day thereafter. In our example, this would mean that the 2007 coupon payment was made on the 17th (Monday) instead of the 15th (Saturday) of September. If the 2007 coupon was calculated as if paid on the 15th of September, this calculation method would be called *unadjusted*. If, however, the 2007 coupon size was based on the period 15 September 2006 to 17 September 2007, this would be called *adjusted coupon*. Apart from weekends, one also needs to regulate how to deal with public holidays, which will differ among countries. In the EUR area, one typically uses the 'TARGET' calendar, which only defines 1st of January, 1st of May, 25th/26th of December, Good Friday and Easter Monday as holidays.

Figure 1.1 (source: Vienna Stock Exchange) shows the price moves of the Austrian government bond in the above example over its life up to 2012. Note that market interest rates were generally falling as a result of the economic downturn from 2008 to 2012, so that the graph shows an upward move in the bond price (the bond now pays a relatively high coupon at 4 %) from 2008. As the bond approaches its maturity in 2016, we expect the traded price to tend to the final principal repayment of 100 % of face value.

Who receives an upcoming coupon payment is determined on the *ex-coupon date*. This is the last day on which an investor buying the bond will receive the next upcoming coupon payment. It is obvious that bonds will sometimes be traded in between coupon payment dates, so that one investor will not receive interest for part of the holding period from the borrower.

Zooming into the graph would not show major jumps around the coupon payment dates (15/09/2006, 17/09/2007, etc.) despite the payment of a coupon. The reason lies in the prices reflecting *clean prices*. If investors sell bonds in between coupon payment dates, they expect to receive interest from the new holder of the bond (buyer) for their hold period since the last coupon payment day. This portion of the coupon is referred to as *accrued interest*. The price at which the bond will be

Fig. 1.1 Price chart of the Austrian government bond as described in this section, 2006–2012

sold is the *dirty price*, which is calculated as clean price + accrued interest. Accrued interest is not produced by traded prices, but simply calculated as the portion of the upcoming coupon that refers to the hold period since the last coupon payment date according to the day-count convention.

1.3 Accrued Interest

In the following we will disregard possible effects of day-count conventions. *Nominal interest rates* are defined as a percentage $R\,\%$ and a time unit to which it is applied, e.g. $4\,\%$ p.a. It is market convention to use one year as time unit when stating nominal interest rates. If a 10-year bond pays a coupon of 4% at the end of each year, this would be preferred by investors over a payment of $10 \times 4\% = 40\%$ at the maturity of the bond, as received coupon payments can be reinvested. Hence, one also has to define the *compounding period* after which interest is paid out. A compounding period of 3, 6 or 12 months results in quarterly, semi-annual or annual interest payments, respectively. If the time unit is the same as the compounding period, the nominal interest rate is also the *effective interest rate* i.

We now let $i^{(m)}$ denote the nominal interest rate p.a. with compounding period $1/m$ years (i.e. compounded m times per year), which leads to the equivalent amount by the end of the year, i.e.

$$1 + i = \left(1 + \frac{i^{(m)}}{m}\right)^m.$$

Correspondingly, $i^{(m)} = m \cdot [(1+i)^{1/m} - 1]$. If we shorten the periods between interest payments further and further, the limit $m \to \infty$ leads to continuous compounding with (nominal) rate

$$r := \lim_{m\to\infty} i^{(m)} = \ln(1+i).$$

Hence, an initial account balance of $B(t_0)$ will give

$$B(t_0 + n) = B(t_0)\, e^{rn}$$

by the end of year n. We can also say that $B(t_0)$ is given by *discounting* the future balance $B(t_0 + n)$ at the continuously compounded rate r, i.e. $B(t_0) = B(t_0 + n) \cdot e^{-rn}$. One can express the dynamics of the continuously compounded bank account by

$$dB(t) = B(t) \cdot r\, dt$$

with initial condition $B(t_0) = B_{t_0}$, and $t_0 \le t \le t_0 + n$. This ordinary differential equation (ODE) can also be extended to the case where r is a deterministic or stochastic function of time (see Chapter 10).

1.4 Floating Rates, Libor and Euribor

Central banks provide a platform for banks to borrow and lend money to each other, which is called *inter-bank market*. The interest rate offered in this market for lending/borrowing is referred to as *Interbank Offered Rate*. Since 1986 the *British Bankers' Association* has been reporting an average of the inter-bank rates used in the London market on a daily basis, and the quoted rate is called *London Interbank Offered Rate* (short: *Libor*). Libor interest rates are published for various maturities, including 1, 3, 6 and 12 months, and we will refer to these rates as Libor1M, Libor3M etc. Note that Libor rates are not only available for British pounds (GBP), but also for many other currencies, including the US dollar (USD), the Euro (EUR) and the Swiss franc (CHF). The inter-bank rates in the EUR-market are compiled by the *European Banking Federation* and quoted as *Euribor* rates.[6]

[6]Concretely, the Euribor rate is determined based on the offering rates of 43 panel banks (as of May 2012), and after eliminating the top and lowest 15 % of the quotes, the Euribor is computed as the arithmetic mean across the remaining figures, rounded to three decimal places.

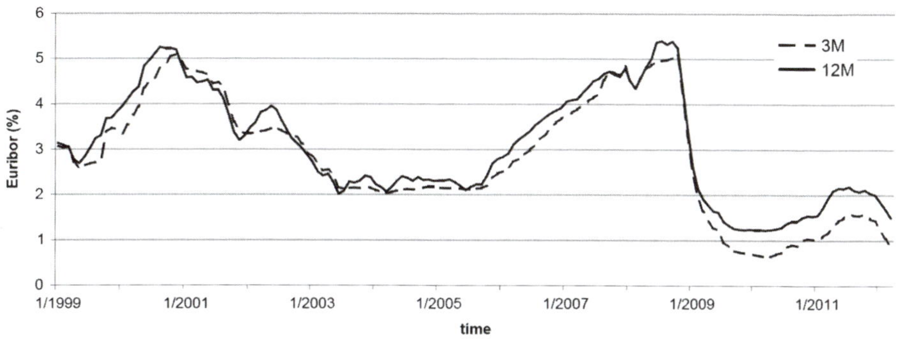

Fig. 1.2 Euribor3M and Euribor12M (01/1999 to 03/2012)

Figure 1.2 depicts the development of the Euribor3M and Euribor12M (in % p.a.) from 1999 to early 2012.[7]

If bank A lends 1mn EUR to bank B for a term of one year, bank B has the obligation to repay the principal of 1mn EUR (principal repayment) plus the interest for the year at Euribor12M. Note that for such an inter-bank loan, the applicable interest rate (here: Euribor12M) will be fixed at the beginning of the period, and not at the end ('*fixing in advance*').

A *vanilla floater*[8] is a variable-interest bond with annual, semi-annual or quarterly coupons. The respective coupon payments, which are paid at the end of every coupon period, are calculated by

$$\text{principal} \times \text{reference interest rate} \times \text{DCF},$$

where DCF is short for *day-count fraction* and describes the coupon period as the proportion of the whole year according to the day-count convention.

Example

Determine the appropriate initial price x of a vanilla Euribor floater issued by a bank which can borrow at Euribor in the markets. Assume a maturity of 10 years, annual coupon payments and a face value of 1.

[7]Source: German Bundesbank, www.bundesbank.de.

[8]Standard products that show no exceptional features are often called '(plain) vanilla', like vanilla ice cream, which seems to be one of the top-selling flavors.

The bond cash flows can be described as follows:

time	investor pays	borrower pays
0	x	
1		Euribor12M (fixed at time 0)
2		Euribor12M (fixed at time 1)
...		...
9		Euribor12M (fixed at time 8)
10		Euribor12M (fixed at time 9)
		plus principal repayment of 1.

Euribor12M (fixed at time 9) reflects the interest rate at at which banks would lend money in the inter-bank market for a year, from time 9 to time 10. The present value[9] at time 9 of the cash flow at time 10 is then 1, and by backward induction one can conclude that the present value of the floater at all coupon payment dates as well as the starting date will equal the face value, so that $x = 1$.

The considerations in the above example lead us to the following observation:

Conclusion

Neglecting credit risk, the value of a vanilla floater equals its face value on its coupon payment dates and on its starting date.[10]

The value of a vanilla floater on its coupon days is simply its face value. In between coupon days, the value of the bond depends on the current market interest rates and the coupon as determined on the last coupon fixing day.

1.5 Bond Yields and the Term Structure of Interest Rates

Suppose that a bond produces known cash flows c_i at times t_i ($i = 1, \dots, N$). Discounting at some fixed intensity y will lead to a *present value* at time t_0 (neglecting day-count conventions) of

$$P(t_0) = \sum_{i=1}^{N} e^{-y\cdot(t_i - t_0)} \cdot c_i .$$

In practice, one will be able to observe the traded market price $P(t_0)$ of e.g. some fixed-coupon bond and the cash flows c_i from the bond at times t_i will be defined in the bond contract. The *market-implied* constant (discounting) intensity y is then

[9]The *present value* is generally defined as the value that a particular stream of future cash flows has at present.

[10]The term 'value' is used here in the sense of fair value. See Chapter 2 for a general discussion.

given by solving the above equation, and y is called the (*continuously compounded*) *yield* of the bond. For given cash flows c_i at times t_i, the mappings

$$P(t_0) \mapsto y \qquad \text{and} \qquad y \mapsto P(t_0)$$

are called *price-to-yield function* and *yield-to-price function*, respectively.

Lemma. *Suppose $t_i > t_0$ and $c_i \geq 0$ $(i \geq 1)$ with $c_i > 0$ for at least one $i \geq 1$. Then every positive market price $P(t_0)$ uniquely determines the continuously compounded yield $y \in \mathbb{R}$.*

Proof. For $y \to \infty$, the present value of the bond tends to 0, and, conversely, for $y \to -\infty$ the present value tends to ∞. As the present value is a continuous function of y, the existence of a solution follows from the Mean Value Theorem and the uniqueness from the monotonicity property of the present value with respect to y. □

Note that for $y = 0$, the present value simply corresponds to the sum of the cash flows. Hence, under the above assumptions we conclude that if the present value is smaller than the sum of the cash flows, the yield y will be positive.

Example (Development of AAA EU Government Yield Curves)

Figure 1.3 depicts the yields of European AAA-rated government bonds as a function of maturity. This representation is often referred to as *yield curve*. In 2005, well before the start of the 2007 Credit Crisis, the yield curve was upward sloping, with yields of around 2 % at the short end, up to approx. 4% at the long end. From the Sep 2008 (just days after the insolvency of Lehman Brothers) curve, it becomes obvious how drastically the shape of the yield curve can change. Short-term yields had increased significantly due to falling demand of short-term investments as investors tried to preserve cash in times of great uncertainty. Finally, as the economic downturn unfolded, a flight to safety alongside with a low short-term interest rates environment led to increased demand for short-term high-quality government bonds, resulting in lower yields, or a *steepening* of the yield curve at the lower end. This is obvious from the Nov 09 and Feb 12 yield curves.

In the above, the yield was determined as the unique discount rate applied to all cash flows of the bond to give its present value. In a slightly different approach, one could understand a bond as a portfolio of different future cash flows. Note that we have previously assumed the interest rate r to be constant across all maturities (i.e. a *flat interest curve*). In practice, however, we will often find interest rates for longer maturities to be higher than for shorter maturities (i.e. a *normal* or *upward sloping* interest curve). We will therefore denote the (continuously compounded) interest rate at time t_0 applied up to time $t_i > t_0$ as $r(t_0, t_i)$.

Keep in mind that interest rates for different maturities can vary greatly. Suppose that the cash flows c_i from a bond at times t_i are known. The present value of the bond (neglecting day count conventions) can also be written as the sum of the cash flows discounted by the interest rates for the respective terms,

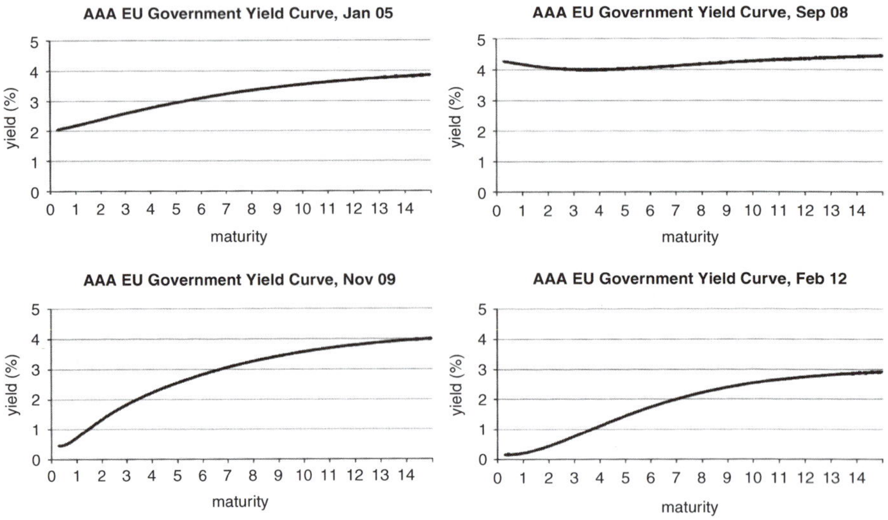

Fig. 1.3 EUR AAA yield curve development 2005 to 2012. Source: European Central Bank

$$P(t_0) = \sum_{i=1}^{N} e^{-r(t_0,t_i)\cdot(t_i-t_0)} \cdot c_i .$$

The yield y will hence be some sort of average over the used discount rates $r(t_0, t_i)$ (or: *zero rates*). Zero rates can be extracted from current bond prices by the bootstrapping method, as described in Section 3.3. The plot of the zero rates as a function of maturity is often called *term structure* or *zero curve*. Chapter 9 will discuss interest rate models in more detail.

1.6 Duration and Convexity

Suppose a currently traded bond price implies a particular yield $y = y_0$. As investors often think in terms of yields, we are now interested to estimate how changes in the yield will change the bond price. Consider the derivative

$$\left.\frac{\partial P(t_0)}{\partial y}\right|_{y=y_0} = -\sum_{i=1}^{N} e^{-y_0\cdot(t_i-t_0)} \cdot c_i \cdot (t_i - t_0).$$

The above expression describes the sensitivity of the bond price, and the following is a widely used sensitivity measure in practice:

Definition. The *Macaulay duration* $D(y_0)$ of a bond with present value $P(t_0)$ and initial yield y_0 is defined as

$$D(y_0) := -\frac{1}{P(t_0)} \left. \frac{\partial P(t_0)}{\partial y} \right|_{y=y_0}.$$

The expression

$$D(y_0) = \sum_{i=1}^{N} (t_i - t_0) \left(\frac{e^{-y_0 \cdot (t_i - t_0)} \cdot c_i}{P(t_0)} \right)$$

makes clear that the Macaulay duration is attained by weighting the contribution of the i-th discounted cash flow to the present value $P(t_0)$ by the time factor $(t_i - t_0)$ (and, conversely, that $D(y_0)$ is a convex combination of the times $(t_i - t_0)$). The Macaulay duration can hence be interpreted as the weighted average cash flow time. For higher yields, later cash flows lose relative weight due to discounting, so that the duration of a cash flow decreases as its yield increases.

Zero-coupon bonds are bonds that do not pay running coupons and only provide one final cash flow at maturity, and their durations are given by their respective maturities.

The sensitivity of the duration to changes in y_0 can be described by the following measure:

Definition. The *convexity* $C(y_0)$ of a bond with price $P(t_0)$ and current yield y_0 is defined as

$$C(y_0) = \frac{1}{P(t_0)} \left. \frac{\partial^2 P(t_0)}{\partial y^2} \right|_{y=y_0}.$$

A Taylor expansion of the present value $P(y)$ at $y = y_0$ gives

$$\frac{\Delta P}{P(t_0)} = -D(y_0) \cdot \Delta y + \frac{1}{2} C(y_0) \cdot (\Delta y)^2 + \cdots,$$

with $\Delta P = P(y_0 + \Delta y) - P(y_0)$. Chapter 13 will further discuss the concept of duration when dealing with the valuation of exotic derivatives. We close the present chapter with an example illustrating the duration/convexity concepts based on a trading strategy.

Example

(**Barbell strategy**) An investor who runs a barbell strategy assembles a portfolio of long and short positions in bonds with different maturities. This is in an attempt to profit from parallel shifts in the yield curve (i.e. yields for all maturities change by (close to) the same Δy, upward or downward).

One can attain market data on bond prices and current yields, and a selection of bonds, each with a face value of 100, could look as follows[11]:

maturity (in years)	3	7	15
y_0	2.58%	3.23%	3.85%
coupon	2.5%	2.25%	4.5%
$P(0)$	99.68	93.64	106.43

All coupons are annual, neglect day-count issues and assume that the first coupon of each bond is paid in a year from now. Verify that the prices and yields as listed above match. Note that the 7-year coupon is larger than the yield, so that $P_{7-\text{year}}(0) < 100$, while the 15-year bond has a coupon in excess of the yield (for exact comparison, you would have to calculate e.g. the equivalent 'continuously compounded' coupon. Why?), so that $P_{15-\text{year}}(0) > 100$. Using the formulas derived in this section, we can compute the durations and convexities of the bonds as

maturity (in years)	3	7	15
duration D	2.92	6.54	11.35
convexity C	8.69	44.55	152.43

Given its long life and its relatively large coupons, the duration of the 15-year bond is significantly lower than its maturity. We can now assemble a portfolio of $x_{3\text{-year}} = 10$, $x_{5\text{-year}} = -10$ and $x_{15\text{-year}} = 2.65$ units of the respective bonds. This is called *barbell* strategy since we buy short-term and long-term bonds, while *short-selling*[12] medium-term bonds (weights at its ends pull the barbell down while you push it up in the middle). Based on the above Taylor approximation, the duration-based change of the portfolio value (all yields change by $\pm\Delta y$) is given by

$$\Delta P_{\text{dur}} = \mp\Delta y \cdot [10 \cdot 99.68 \cdot 2.92 - 10 \cdot 93.63 \cdot 6.54 + 2.65 \cdot 106.43 \cdot 11.35] = 0.$$

The convexity-based change of the portfolio value, on the other hand, is positive for both negative and positive changes to the yield, which is mainly driven by the large convexity of the long-dated 15-year bond:

$$\Delta P_{\text{con}} = \frac{(\pm\Delta y)^2}{2} \cdot [10 \cdot 99.68 \cdot 8.7 - 10 \cdot 93.63 \cdot 44.55 + 2.65 \cdot 106.43 \cdot 152.43] = \frac{(\Delta y)^2}{2} \cdot 9{,}935.$$

Hence, judging by a 2nd-order Taylor approximation, if all yields *widened* by 1 %, the portfolio value would rise by 0.5, and if all yields fell by 1 %, the portfolio value would rise by 0.5 as well, so that we profit from parallel yield curve shifts in either direction. Looking at the yield curve developments in Figure 1.3, where would you see the major risk in implementing such a strategy?

[11]The yield/price quotes used here roughly correspond to EU AAA government bonds as of Jan 2005 (cf. Figure 1.3. Yield/price quotes for government bonds can e.g. be obtained at www.bloomberg.com/markets/.

[12]Short-selling can be imagined as borrowing today's price of a stock, while the repayment will be again at the (future) price of the stock. If the stock price falls, the short-seller will gain, as he has to repay less.

1.7 Key Takeaways, References and Exercises

Key Takeaways

After working through this chapter you should understand and be able to explain the following terms and concepts:

- Nominal interest rates, annual/semi-annual/quarterly/continuous compounding
- Day-count conventions (30/360, Actual/365, Actual/360), Business-day conventions (TARGET)
- Bond prices typically rise/fall as market interest rates fall/rise, and tend to face value as maturity is approached
- Fixed-rate vs. floating rate bonds (with Libor, Euribor as reference rate-coupons are 'fixed-in-advance')
- Bond price as function of the bond yield vs. bond price as function of the cash flows discounted at the zero rates
- Yield curves: flat, normal, shape change over time
- Duration/convexity: definitions, link to Taylor approximation of value change, barbell strategy

References

Well-structured and comprehensive discussions of the topics covered in this section can be found, for example, in Hull [41] or Wilmott [75]. Current and historical interest curves can be viewed at websites of exchanges, such as www.deutsche-boerse.com, www.swx.com or www.wienerborse.at, or from central banks including www.bundesbank.de, www.snb.ch and www.ecb.int.

Exercises

1. Calculate the point in time at which some initial capital c has doubled, if interest is compounded (i) annually, (ii) monthly or (iii) continuously, using an interest rate of R % (p.a.). In particular, give a numerical answer to the above for $R = 5$.
2. A generous benefactor launches a foundation that will award an annual prize for extraordinary accomplishments in the field of mathematics, similar to the Nobel Prize. Assume interest can be earned at 4 % p.a. and compute the required initial capital c such that 1mn EUR can be awarded to the respective laureate each year (i) for 10 years, (ii) for 100 years, or (iii) forever.
3. In addition to the Macaulay duration, the *modified duration* is widely used. It also measures the sensitivity of the present value of a future cash flow stream with respect to the discounting rate, but assumes discrete (typically annual) interest payments and uses the yield-to-price function $(1 + y_m)^{-(t_i - t_0)}$ instead of $\exp(-y(t_i - t_0))$. Derive an explicit formula for the resulting modified duration.

Exercises with Mathematica and UnRisk

4. (a) Use the commands `MakeFixedRateBond`, `CashFlows` and `Valuate` to determine the exact dates and amounts of the cash flows of the government bond described in Section 1.2 (with the ACT/ACT day-count convention).

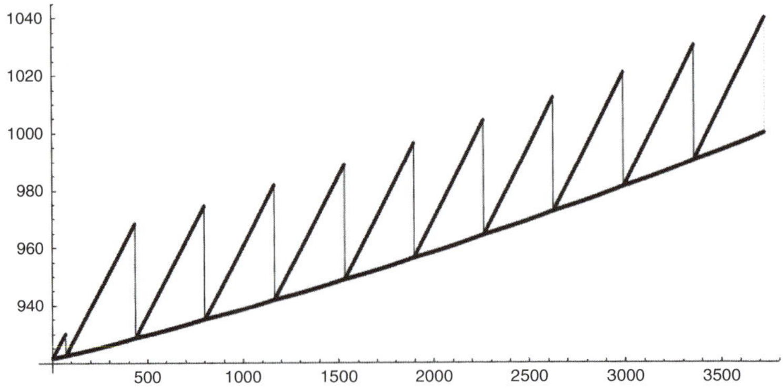

Fig. 1.4 Dirty and clean price with constant annual interest rate of 5 %

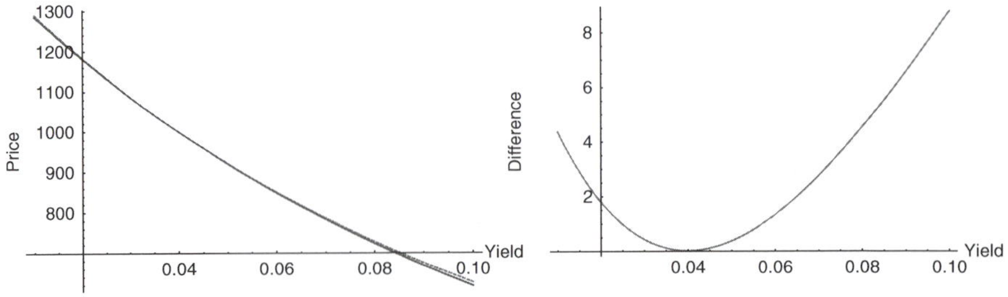

Fig. 1.5 Yield-to-price function of the government bond (Section 1.2) and the zero-coupon bond (left), and the difference between the two functions (right)

(b) Use the command `MakeYieldCurve` to plot the 'dirty' and the 'clean' price of this bond as a function of time up to maturity, under the assumption of a constant interest rate of 5 % (see Figure 1.4).

(c) Test the sensitivity of these curves as the interest rate is changed to 4 % or 6 %. Implement a scroll bar to change the interest rate.

(d) Test how the curves change if the day-count convention 30/360 is used.

(e) Assume that the zero rates follow the law

$$r(2006 + t_0; T) = \frac{2 + 3\exp(-t_0/5)}{100}$$

from 2006 onwards, but are constant for $2006 + t_0$. How do the plots of the dirty and the clean price of part (b) change under these new assumptions?[13]

5. Suppose $y = 0.04$. Use UnRisk to construct the zero-coupon bond by choosing the nominal amount and the maturity, such that the bond has the same price, yield and duration as the government bond in Section 1.2. Assume an ACT/ACT day-count convention. Illustrate that the convexity of the two bonds is different. Plot the yield-to-price functions for $y \in [0.01, 0.1]$ (see Figure 1.5).

[13]The forward interest rates as implicitly used here will be discussed further in Chapter 9.

Financial Products 2

2.1 Bonds, Stocks and Commodities

Bonds

In Chapter 1, bonds have been introduced as an important class of financial assets which is structurally similar to loans. The authorized *issuer* promises in the bond contract to make future payments according to a fixed schedule, up to some final time T (the *term* or *maturity* of the bond).[1] The promised payments typically consist of the principal (or: face value) of the bond (e.g. 10,000 EUR) at time T and a regular (for example, annual, semi-annual or quarterly) *coupon* (e.g. 500 EUR at the end of every year). If no coupon is paid, there is only one payment at maturity (typically after one year or less) and the bond is called *zero-coupon bond*. Coupon payments can be an initially fixed amount, e.g. 5% p.a. of the principal. Alternatively, the size of the coupon can be linked to some reference interest rate, e.g. Libor+1% (see Section 1.4). If the principal is paid in one lump sum at maturity, the bond is called *bullet*. Otherwise one speaks of an *amortizing bond*.

Note that the issuer will often hold an auction when initially selling the bond to investors. The initial price of the bond is determined by the bids of the investors, and can be different from the face value. Given a face value of 100, if investors offer more than 100, the bond is said to sell at a *premium to par*. Conversely, if investors offer less than 100, the bond sells at a *discount to par*. Once the bond is sold to the initial investors in the primary market, these investors might decide to sell the bond to other parties in the secondary market. Bonds are *debt securities* and can easily be traded privately (for example, through bond funds, insurance companies or banks), or exchanges might provide a platform to match buyers and sellers. Note that a bond investor will record the bond as an asset on its balance sheet, while the issuer will report it as a liability (i.e. as an obligation to pay money in the future).

[1] Due to their fixed payment schedule, bonds are also referred to as *fixed income* products.

H. Albrecher et al., *Introduction to Quantitative Methods for Financial Markets*, Compact Textbooks in Mathematics, DOI 10.1007/978-3-0348-0519-3_2,

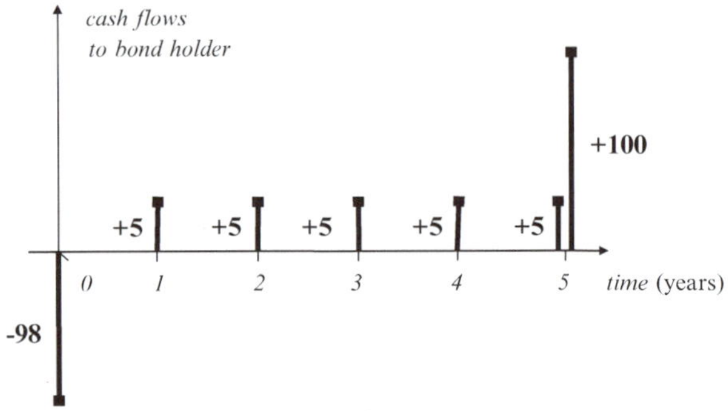

Fig. 2.1 Cash flows to the bond investor: 5-year bullet, face value 100, 5% annual coupon and initial price 98

Stocks

A *stock* (or: share) represents capital paid into a company in return for ownership, either by the initial founders or at a later stage. A stock is a security that gives its holder a number of rights, including

- the right to receive dividends;
- the right to participate, speak and vote at General Meetings;[2]
- the right to receive new shares. As additional share capital is raised, this will typically be first offered to current shareholders so that their voting power is not necessarily *diluted*;
- the right to participate in the distribution of liquidation proceeds once all other liabilities have been repaid in full.

Note that stocks can also be held and traded privately, they are not necessarily *listed* at stock exchanges. Listed companies might have a large *free float*, i.e. a large portion of their stocks is owned by many different equity investors, which provides sufficient liquidity for almost continuous trading. Many regulators require larger holdings of shares of a company to be (publicly) disclosed (e.g. UK: once the holding exceeds 3% of the number of outstanding shares).[3,4]

Listed companies are required to publish detailed information in the form of quarterly and annual reports. Information rules can be imposed by the regulator or the respective stock exchange, and might differ from market to market.

[2]A stock company is required by law to hold Annual General Meetings where past and future activities are discussed, fiscal information is reviewed and the Board of Directors is elected.

[3]Larger strategic holdings by long-term investors are not counted into the free float, together with government holdings or holdings of founding investors.

[4]Stock prices of otherwise comparable companies with only a small free float can be more volatile. Some hedge funds had to experience this in 2008, as they lost more than 20bn GBP when closing short positions on Volkswagen stocks. Porsche had just announced that it had acquired as much as 74% of Volkswagen stocks. Only a relatively small portion of stocks was still free-floating, so that prices sky-rocketed within hours due to the sudden demand from hedge funds and the limited supply.

Stock Indices

To describe the performance of an entire stock market, for example a selection of companies listed at the Frankfurt stock exchange, *stock indices* are computed and published and can be tracked over time. A stock index is a linear combination of a set of stock prices and is published by the stock exchange itself (e.g. DAX (Frankfurt), DJIA (New York), Nikkei (Tokyo), SMI (Zurich)) or by information providers (e.g. S&P 500 (500 large cap stocks traded in the US), the Dow Jones Industrial Average (short: DJIA, 30 large US based companies that are publicly traded)).

Suppose an index contains n stocks with stock prices $s_1, s_2, \ldots, s_n$ and numbers of outstanding shares $nos_1, nos_2, \ldots, nos_n$. The *market capital* mc_i of stock i is simply its current stock price times the number of its outstanding shares, i.e. $mc_i = s_i \cdot nos_i$. Indices can then be calculated as *price-weighted indices* or *market-value-weighted indices*. A price-weighted index I_p is calculated as

$$I_p = \frac{\sum_{i=1}^{n} s_i}{\text{number of stocks (adjusted for splits)}},$$

and it will become clear from the example below how the number of stocks (adjusted for splits) is computed. A market-value-weighted index I_m, on the other hand, is calculated as

$$I_m = c \cdot \sum_{i=1}^{n} mc_i = c \cdot \sum_{i=1}^{n} s_i \cdot nos_i$$

for some constant $c > 0$. Clearly, a market-value-weighted index can move with only a small number of large companies that have large market capital, while small market-capital companies have relatively more weight in a price-weighted index.

Now assume a company decides to split its stocks so that current owners receive k new stocks for every stock they own. If a stock trades at 33 GBP before the split, the new stocks just after a 1:3 split will trade at 11 GBP and each investor will hold three times as many shares as before. Stock splits have no effect on market-value weighted indices since $s_i \cdot nos_i = \frac{s_i}{k} \cdot (nos_i \cdot k)$. To understand the effect of stock splits on price-weighted indices, consider the following example.

Example (Downward bias of price-weighted indices)

Consider a price-weighted index on a set of two stocks A and B. At time t_0, stock A trades at 100 EUR per share and stock B at 40 EUR. At some later time t_1, stock A rises to 200 EUR and stock B to 50 EUR. We calculate $I_0 = \frac{100+40}{2} = 70$ and $I_1 = \frac{200+50}{2} = 125$. Company A decides that its stock trades too high and splits it 1:2. After the split, at time t_1^+, the number of stocks has to be adjusted from 2 to *number of stocks*$_{\text{adj}}(t_1^+)$, so that the index does not change. Hence, one solves

$$\frac{200/2 + 50}{\text{number of stocks}_{\text{adj}}(t_1^+)} = \frac{200 + 50}{2}$$

to find number of $\text{stocks}_{\text{adj}}(t_1^+) = 1.2$. At some later time t_2, A and B trade at 118 EUR and 50 EUR, respectively. The index will now be $I_2 = \frac{118+50}{1.2} = 140$. Note that A performed relatively better than B over the period $[t_1, t_2]$. Without the stock split, the index would have been $I_2^{\text{no split}} = \frac{236+50}{2} = 143$. Due to the split, stock A has lost some influence on the index. This effect is known as *downward bias of price-weighted indices*, because successful companies are more likely to perform stock splits when their stock price keeps rising.

Without going into further detail, keep in mind that different ways of computing indices measure market performance differently. Also, some indices are published both as *price performance indices* and *total return indices*, depending on whether dividend payments are included. Finally note that indices have become a fundamental tool of well-developed financial markets, as they allow to assess the performance of single assets relative to an entire market, to evaluate relationships between financial or economic variables and market performance, to construct index portfolios tracking the overall market, and to hedge against adverse (sub-)market movements through index-based derivatives.

Currencies (FX)

Currency or foreign exchange (short: FX) markets provide a platform for trading currencies. Currencies are traded directly between two parties over-the-counter (short: OTC), without going through an exchange, and most trades are between banks. A particular trade consists of a currency *pair*, such as EUR/USD, USD/JPY, AUD/USD, or USD/CHF. A *market maker* could quote EUR/USD 1.2938/1.2940. EUR would be the *base currency*, as the quotes refer to 1 EUR, and USD the *quoted currency*.[5] The quote is given as *bid/ask*, i.e. the market maker would buy 1 EUR for 1.2938 USD, and sell 1 EUR for 1.2940. The difference between the two quotes is called *bid-ask spread*. Currencies are typically traded in contract sizes (or: *lot sizes*) of 100,000 units of the base currency, but smaller sizes are also offered to retail clients. The FX market is one of the largest markets if measured by transaction volume. The average daily turnover in April 2010 was 4,000bn, which marked a 20 % increase over the April 2007 figure (cf. BIS [74]). FX rates can be very volatile and Figure 2.2[6] depicts the development of the EUR-USD exchange rate from 1999-2012.

Commodities

Commodities, such as oil (different types), gas, coal, electricity, base metals, precious metals, agricultural goods (soy, wheat, corn, pork bellies) or soft commodities (coffee, cocoa, sugar, cotton, orange juice), can be traded in the *spot market* or the *forward/future market*. Upon trades in the spot market, the buyer receives control over the traded good immediately or at the latest within a short *settlement period*.

[5]Which currency in a traded pair is quoted as base currency is mostly based on historical convention.

[6]Source: www.bundesbank.de

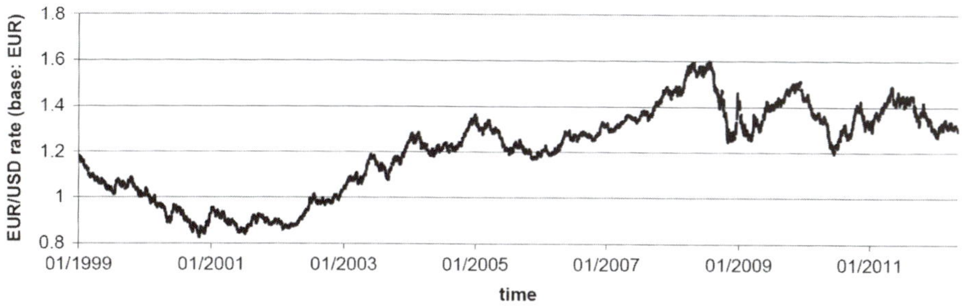

Fig. 2.2 Historical EUR/USD exchange rates (base: EUR) 01/1999-05/2012

The bulk of the trades are however executed in the forward market. For example, when entering a contract in the forward market, one counterparty might accept the obligation of delivering 10 megawatt-hours of electricity per hour throughout some future month. The other counterparty then has the commitment to buy this quantity of electricity at the scheduled times at a price fixed today. We will further discuss this kind of contracts in Section 2.3.

2.2 Derivatives

Financial instruments whose value depends on the price of some other *underlying* product are called *derivative instruments* (short: *derivatives*).[7]

Derivatives that give the right (but not the obligation) to engage in a financial transaction at a later point in time are called *options*. An example of an option would be the right to buy or sell an asset at some later time T at a price fixed today. The analysis of such *contingent claims* is one of the main fields of modern financial mathematics.

Derivatives can be standardized contracts that are traded at stock exchanges, or they can come in the form of products tailored specifically to the requirements of the counterparties. Such non-standard contracts are typically traded over-the-counter (OTC).

Why are derivatives traded and who would have particular interest in entering into derivative contracts? Two possible motivations for engaging in the derivatives market are listed below:

- **Hedging:** Consider the following example. An exporting company, which produces a machine in Europe, has agreed to sell this machine upon completion to a client in the US at a fixed USD amount. Assume that the production costs of this machine will mainly incur in EUR. The company is therefore exposed

[7]Note that the underlying of a derivative contract can again be a derivative with respect to another underlying, and so on.

to currency exchange rate risk between the time of production and the time of the sale. An unfavorable move of the EUR/USD rate (i.e. that the USD loses value compared to the EUR) will lower the company's profit. The company can now partly or fully mitigate this risk by entering into an FX forward contract. This contract fixes the future exchange rate at a certain level. Mitigating risk by taking on a portfolio of one or more financial instruments[8] is called *hedging*. In particular, note that the exchange rate risk is now borne by the counterparty in the FX forward contract (which will often be a bank) rather than by the company or the buyer of the machine.

- **Taking uncovered positions:** Market participants can also take a position in a derivative without being in some way exposed to the underlying risk. This would be called *taking an uncovered position*, and it can lead to a profit if a particular market view proves true. For example, one could take the position of the counterparty in the above FX forward contract thinking that the USD will gain value against the EUR. If the USD then actually appreciates versus the EUR, this position will bring a profit. Taking positions in derivative products typically allows for more specific and efficient trading strategies than those realizable by holding positions in only the underlyings themselves (cf. Section 2.5).[9]

2.3 Forwards and Futures

In the spot market, goods and payments are exchanged (e.g. domestic against foreign currency, cash against stocks, cash against copper etc.) immediately or at the latest within a short settlement period. Conversely, it can be agreed to execute the exchange at some later time. If the later exchange is unconditional, this contract type is called *forward* contract. Concretely, a forward contract defines the obligation to trade a good (e.g. a stock) at some time T at an agreed price F. The buyer of the underlying is said to have a *long position* in the forward, and the seller has a *short position*. The transaction (the payment of the forward price and the delivery of the good) will be executed at time T. If the price S_T of the underlying at time T is larger than F, then the contract has the value $S_T - F > 0$ to the buyer. Conversely, the seller has to sell below market, and therefore takes a loss of $F - S_T$. The pay-offs

[8] In our above example, the hedging portfolio consists of one FX forward contract.

[9] Note that we often take views when making financial decisions. For example, when part-financing the purchase of a house through a bank loan, the borrower might be able to choose between fixed or floating interest rates, or to fix an upper interest rate limit (also: *cap*) in the case of floating interest rates. It also used to be popular to finance real estate by loans in foreign currencies with lower borrowing rates, for instance, financing a German house with a CHF loan when interest rates in CHF were lower than in EUR. During the economic downturn starting in 2007, however, the CHF greatly appreciated in value against the EUR, so that CHF-denominated liabilities required a significantly higher EUR amount to be repaid. Even when choosing a mobile phone contract, one will usually decide on a particular contract duration/fee combination and hence take a view on phone contract terms in e.g. 12 months from now.

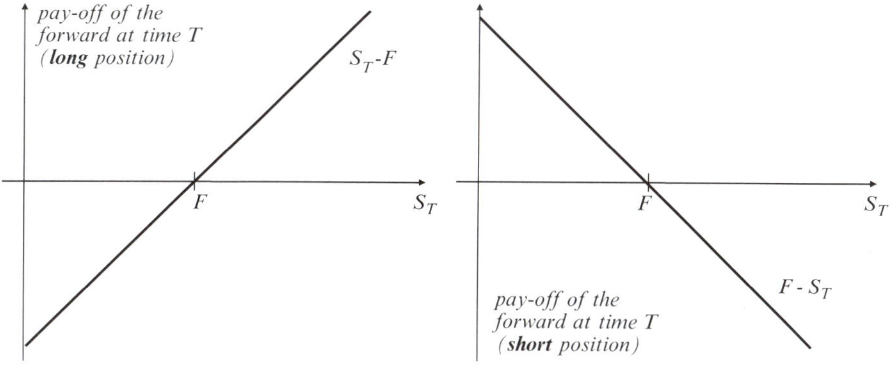

Fig. 2.3 Pay-off of a long/short forward contract at maturity T

of the long and short forward contract are depicted in Figure 2.3. Note that only the short position faces a potentially unbounded loss.

Forwards are not only traded on underlying stocks, but also on interest rate products, other financial instruments, and commodities. The *standardized* version (in terms of the quality of the underlying, the maturity, the contract size, etc.) of the OTC-traded forwards are called *futures*. Futures are traded at futures exchanges. The standardized nature of futures makes it easier to take a counter-position to close a certain position (e.g. closing a long position by adding a short position – netting off the two pay-offs in Figure 2.3 gives then zero) and ensures increased trading liquidity. Futures exchanges include the Chicago Mercantile Exchange (www.cmegroup.com), the Intercontinental Exchange Inc. (www.theice.com) and the European Energy Exchange in Leipzig (www.eex.com).

Finally note that, in practice, instead of *physical settlement* (i.e. the underlying will be physically delivered against the payment of the futures price at maturity), most future contracts will be *cash settled* (i.e. one party will receive a payment corresponding to the value of the contract at the time of closing the position). The actual financial settlement of future contracts will be done through a *clearing house* as central counterparty.[10] As future contracts can have a maturity of up to several years, the price of the underlying in the spot markets (and hence the value of the of the futures contract) can fluctuate significantly up to maturity of the future contract. Pricing of futures and lowering the risk of the futures counterparty not fulfilling its obligations under the contract will be further discussed in Section 3.2.

[10]Currently (2012) LCH.Clearnet (www.lchclearnet.com) is the largest clearing house for derivatives.

2.4 Swaps

Swaps are contracts between two counterparties to exchange two cash flow streams. Consider the following example of a *fixed-for-floating* interest rate swap.

Example (10-year vanilla interest rate swap)

Effective/Termination date: 25 April 2012/25 April 2022
Notional amount: 8,000,000 EUR
Party A pays and party B receives: quarterly Euribor3M, fixing in advance (ACT/360)
Party B pays and party A receives: 2.320 % p.a., paid annually, (30/360).

The party in an interest rate swap which pays the fixed rate is called *fixed rate payer*. In the above example, counterparty A is the *fixed rate receiver*. Cash flows under the swap (from A to B, and vice versa) are calculated by applying the respective interest rates to the *notional* amount, which is similar to the principal of a bond. However, the notional itself is actually never exchanged between the parties. Note that arbitrary reference interest rates can be used when defining a swap, however, for Euribor/Libor common rates include 1M, 3M, 6M or 12M. The two different cash flow streams in a swap are referred to as *legs*. The floating Euribor3M cash flow in the above example would be called *floating leg*, the cash flow linked to the fixed interest rate *fixed leg*. Even for more complex swap products, one leg will typically have a plain vanilla structure as above, while the structure of the other leg may be more complex. From a certain degree of complexity upwards, the contracts are called *structured swaps* and will be further discussed in Chapter 13.

Swaps are typically tailored to the needs of at least one of the counterparties and hence traded OTC. It has become an industry standard to document a swap contract based on a *swap master agreement* as developed by the *International Swaps and Derivatives Association*[11]. Using standard documentation and standard contract terms considerably lowers documentation risk and legal risk, and allows to compare different contracts more easily.

The value of a swap (from the viewpoint of the respective counterparty, A or B) typically changes over its life as market conditions (e.g. interest rate levels) change. If the swap has value to e.g. party A, A bears the risk that party B will not be able or willing to entirely fulfil the contract. Hence, A might contractually require B to post some sort of *collateral* (e.g. cash or government bonds) to cover this risk. Initially, the fixed rates in the case of vanilla interest rate swaps are mostly set such that the swap has zero value at the beginning (and this fixed rate is referred to as *swap rate*). If the swap in the above example had had zero value on the 25th of April 2012, the 10-year EUR-swap rate would have been 2.320 % then. Note that plain vanilla swaps are very liquid instruments, which is partly due to the standard

[11] ISDA, www.isda.org

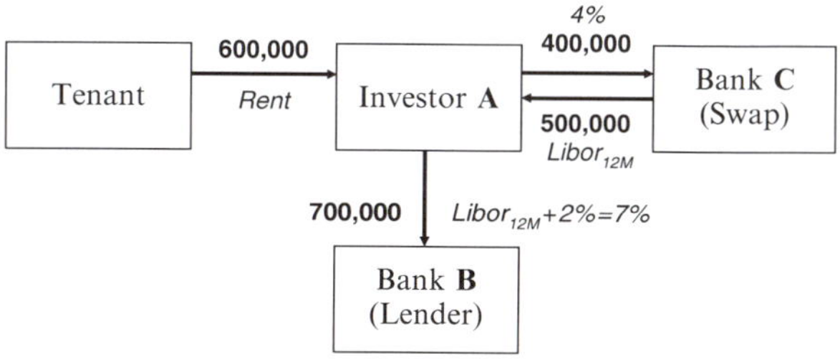

Fig. 2.4 Cash flows on a loan interest payment day for the example below

definitions of the ISDA documentation and publicly available benchmark quotes (for example, ISDAFIX). In general, swap contracts can also have non-zero initial value, so that one counterparty would make an initial payment to the other counterparty. Similarly, one can choose a structure where the notional increases or decreases over time (*accretive principal swap* or *amortizing swap*, respectively), such that swap contracts can be tailored for managing interest rate risk arising from specific loans or bonds. We close this section with an example of how swaps can be applied to the hedging of interest rate risk.

Example (Interest rate hedging)

Suppose A is a real estate investor and buys a building for 12 mn EUR that produces 600,000 EUR in net rental income every year. A only has 2 mn EUR in cash and borrows the remaining 10 mn EUR from bank B for a term of 7 years and at an interest rate of Libor12M+2%. As the rental income from the tenant is fixed in the lease contract, there is the risk that Libor12M rises very high, so that the interest payment to B cannot be covered from the net rental income any longer. To mitigate this risk, bank B asks A to enter into a fixed-for-floating interest rate swap contract. Another bank C offers to pay Libor12M against a fixed rate of 4 % paid by A (assume yearly payments). The notional is set at 10 mn EUR and the termination date is in 7 years from now. Figure 2.4 shows the cash flows on a loan interest payment day if Libor12M= 5 % on some fixing day. Note that A can only cover the interest due to the extra payment from the swap counterparty. Conversely, if LIBOR12M was below 4% on a fixing day, A would have to pay 10 million×(4%−LIBOR12M) to the swap counterparty C (in which case having a swap in place would be a disadvantage for A). A has effectively locked in its interest plus swap costs at 4%.

2.5 Options

In the financial context, an *option* is the right, but not the obligation, to purchase or sell some underlying asset (e.g. a stock) at some time $T \geq 0$ at a pre-defined price K. The price K is called *strike price* (or simply: *strike*) and T is called *expiration date* (or: *expiry*). One distinguishes between *call options*, which give the option buyer the right to buy, and *put options*, which are rights to sell (to 'put an

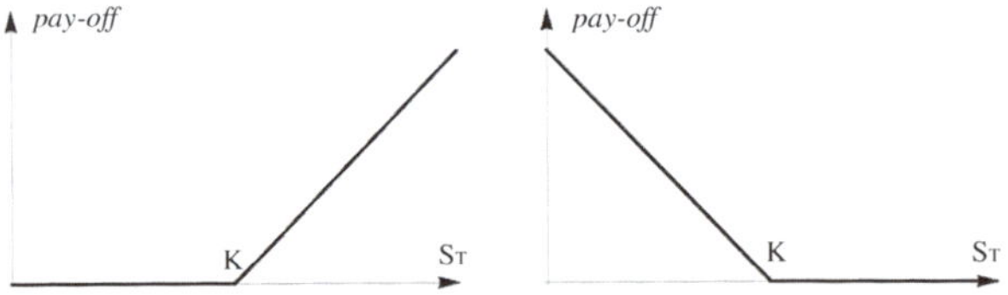

Fig. 2.5 Pay-off of a call (left) and a put (right) option with strike price K, as function of the stock price S_T

asset on the market').[12] The buyer of an option is said to have a *long position* in the option, while the seller has a *short position*.

The *pay-off* of an option is its value at the time of its exercise. In the case of a call option with strike K on an underlying stock with price S_T at expiry T, the pay-off C_T is given by $S_T - K$ if $S_T > K$, and 0 if $S_T \leq K$. In the latter case the stock can be purchased at a price lower than K in the market, and hence the option will not be exercised. Altogether, one can write

$$C_T = \max(S_T - K, 0) = (S_T - K)^+.$$

Similarly, the pay-off P_T of a put option is given by $P_T = (K - S_T)^+$ (see Figure 2.5).

Example (Leverage effect of options)

Let $S_0 = 100$ EUR be the price of a stock today, and let some call option on the stock have strike $K = 120$ EUR, expiry T and initial price $C_0 = 5$ EUR. How can one profit, if the stock price will rise significantly until T?

(a) Buy the stock today at $S_0 = 100$. If it turns out that $S_T = 130$ EUR, the stock holder will have made a 30 % profit on the investment over the period $[0, T]$.

(b) Alternatively, you could buy the call option today. If $S_T = 130$ EUR, the option will be exercised and the stock can be attained at time T at 120 EUR. If the stock is then immediately sold in the market, this would give a profit of $\frac{130-120-5}{5} = 100\%$ on the investment.[13]

The increased percentage profitability of buying the option compared to buying the underlying stock is called *leverage effect*. Note, however, that strategy (b) also bears the risk of receiving zero pay-off (if $S_T < 120$), so that the entire investment would be lost in that case. Similarly, one can profit from falling stock prices in a leveraged structure by buying put options.

So far we have only considered the possibility of the options being exercised on one specific date, the expiry date. Such options are called *European options*. Other types of options are also offered in the market. For example, *American options* can

[12]Calls were first traded as standardized contracts at the CBOE (Chicago Board Options Exchange) in 1973, and puts followed in 1977. Today options are traded at more than 50 exchanges worldwide. The most important European options exchanges include EUREX (www.eurexchange.com) and LIFFE (www.liffe-commodities.com).

[13]In practice, the option holder will typically receive a *cash settlement* of $130 - 120 = 10$ EUR, instead of receiving the stock physically and paying 120 EUR.

be exercised at any point in time up to expiry, or *Bermudan options* can be exercised at pre-defined discrete times up to expiry. Note that options can deviate from the *plain vanilla* structure as explained here. Such more complex options are referred to as *exotic options*, and are traded OTC. Examples of exotic options include:

- *Asian options:* the stock can be sold at expiry at the average stock price up to expiry (or, in a slightly different structure, the strike is fixed and the pay-off is given as the difference between the average stock price and the strike if this difference is greater than 0, and 0 otherwise). The price averaging dampens the effect of highs and lows in the price development of the underlying.
- *Barrier options*: in this case, the pay-off of this otherwise European option depends on whether the stock price crosses a certain barrier up to expiry. For the so-called *knock-out option*, the option is canceled (i.e. the pay-off becomes 0) as soon as the defined barrier is crossed, for the *knock-in* version, the European pay-off is only made if the barrier has been crossed.[14]
- *Compound Options*: are options on options.
- *Digital Options*: have the constant pay-off 1, in case the stock price S_T exceeds the strike K at expiry, and 0 otherwise (in the case of a call).

This list could be arbitrarily extended, in particular for the remaining 22 letters of the alphabet.

2.6 Key Takeaways, References and Exercises

Key Takeaways

After working through this chapter you should understand and be able to explain the following terms and concepts:

- Bond, bond issuer, zero-coupon bonds, bullet, principal/face value, premium/discount to par
- The rights of a stock holder
- Market-value-weighted vs. price-weighted stock indices, the downward bias
- In the context of FX, bid/ask quotes, bid/ask spread, base currency
- The difference between forwards and futures
- Swap contracts
- European, American, Asian, Bermudan and Barrier options, and the leverage effect of options

[14]Barrier options are amongst the most liquid OTC options and are an important building block of many structured products (cf. Chapter 13).

References

Details and calculation methods for stock indices at the Vienna stock exchange can be found at www.indices.cc/indices/, for the DAX and related indices see deutsche-boerse.com and for information on indices of the Swiss stock exchange www.six-swiss-exchange.com/trading/products/indices_en.html. Other global index providers include FTSE (www.ftse.com/indices/) and MSCI (www.msci.com/products/indices/). For a detailed discussion of financial instruments and their relevance in practice, consult e.g. Wilmott [75].

Exercises

1. What is the *number of outstanding shares* (NOS) of the Swiss company Asea Brown Boveri (ABB)? At what stock exchanges are ABB stocks *listed*? Plot the price development of ABB stocks over the last 5 years.
2. What stocks does the Dow Jones Industrial Average (DJIA) consist of? What is the composition of the DAX? How is the ATX calculated?
3. Check and list the contract specifications of various PHELIX futures as traded at the European Energy Exchange.
4. What are the current prices of European options on the S&P500 index as listed by the CBOE?
5. (a) Explain the difference between holding a long position in a forward contract with a forward price of 50 EUR, or a long position in a call option with strike 50 EUR.
   (b) A trader expects a stock price to rise and would like to profit in case his view proves true. The current stock price is 29 EUR and a European call option ($T = 3$ months, $K = 30$ EUR) prices at 2.90 EUR. The trader can invest a total of 5,800 EUR. Identify two strategies – investing in the stock, or taking a long position in the call options. Specify the absolute and relative (percentage) profit/loss of the two strategies, depending on the stock price in 3 months from now.
6. A company has information that it will receive a certain amount in foreign currency in 4 months from now. How can you hedge this transaction using (i) a forward contract, or (ii) an option contract. What will the structural difference between (i) and (ii) be?
7. Search the internet to find out what types of Asian options are commonly used.
8. (a) Describe the pay-off of the following portfolio: a long position in a forward contract on a stock and a long position in a European put option, both with expiry T. The strike K of the option shall equal the fair forward price of the stock at time 0.
   (b) Is the following statement true? Explain your answer.
   'A long position in a forward contract is equivalent to a long position in a European call option and a short position in a European put option.'

The No-Arbitrage Principle

3

3.1 Introduction

The term *arbitrage* is used for making *risk-free* profit by buying and selling financial assets in one's own account. Let π_t be the value of a portfolio at times $t \geq 0$, with $\pi_0 = 0$. An *arbitrage strategy* is then formally described as

$$\mathbb{P}(\pi_t \geq 0) = 1 \quad \text{and} \quad \mathbb{P}(\pi_t > 0) > 0 \quad \text{for some } t \geq 0.$$

It is natural to define that the price of an instrument is *fair*, if adding it to the market does not produce arbitrage opportunities.[1] Consider the following simple example of *cross-market arbitrage*.

Example

Assume that a stock trades both in Chicago and in Frankfurt. The current stock price is 100 USD in Chicago and 70 EUR in Frankfurt. The EUR/USD exchange is currently 1.33 (EUR base). Neglecting transaction costs, this would imply an arbitrage opportunity as follows:

- Buy 100 stocks in Frankfurt.
- Immediately sell the stocks in Chicago.
- Exchange the so-attained USD amount into EUR.

The resulting risk-free profit is

$$100 \cdot \left(\frac{100}{1.33} - 70 \right) \text{EUR} = 519\,\text{EUR}.$$

Due to market transparency, opportunities of arbitrage like the above only exist for very short time periods. If many market participants implemented the strategy in the

[1] In particular, under the assumption of no-arbitrage, goods that produce the same cash flows over time will be required to have the same price ('*law of one price*').

H. Albrecher et al., *Introduction to Quantitative Methods for Financial Markets*,
Compact Textbooks in Mathematics, DOI 10.1007/978-3-0348-0519-3_3,

above example, the increased demand for the stock in the Frankfurt market would increase the Frankfurt price, while the additional supply of stocks in the Chicago market would lower the price there, so that the arbitrage opportunity would quickly disappear.

Market participants that exclusively work on exploiting arbitrage opportunities are called *arbitrageurs*. The presence of such arbitrageurs ensures that arbitrage opportunities disappear rapidly once discovered.[2] When analyzing financial markets, it is hence commonly assumed that arbitrage opportunities do not exist (sustainably). In particular, derivative instruments will be priced in such a way that no arbitrage opportunities arise by adding the derivative to the market. This consideration is fundamental to modern pricing theory for financial markets and is often referred to as the *no-arbitrage principle* (see exercises 1–4).

The following assumptions are widely used when modeling (idealized) financial markets:

- **There do not exist any arbitrage opportunities.**
- **Lending and borrowing rates are equal:** funds can be lent and borrowed at the same interest rate. Usually this assumption is sufficiently satisfied for banks of good creditworthiness during bull markets. During economic downturns, however, banks might find it more expensive to borrow funds due to a drop in supply, so that borrowing rates will turn out higher than lending rates for most participants.
- **No transaction costs:** in practice, the buying and selling of financial instruments will produce transaction costs (fees to exchanges, broker commissions etc.). Still, these costs will often be negligible for large market participants, so that throughout this book we will assume for simplicity that transaction costs do not play a role.[3]
- **Short-sales are allowed:** the term *short-selling* describes a procedure that allows to sell an asset today at today's price while only having to physically deliver it at some later time, i.e. to take a *short position* in the asset. In practice, several issues have to be addressed for short sales, for example how to deal with dividend payments. In principle, large market participants can easily enter into short-sale contracts, but tighter regulation of short sales has been a much discussed topic recently.[4]
- **Financial assets can be split arbitrarily:** one can buy or sell arbitrary (also non-integer) numbers of assets.

---

[2]Modern means of communication and real-time price systems have significantly improved market transparency.

[3]This assumption will have to be reconsidered for certain markets, such as commodity markets. For example, shipping and insurance costs can be significant, so that prices between different market places can differ significantly without implying opportunities of arbitrage.

[4]For further details, check the current EU short sale regulations at ec.europa.eu/internal_market/securities/short_selling_en.htm

- **No dividend payments:** in the following we will assume that no dividend payments are made unless stated otherwise. This assumption is not fundamental, but improves the readability of the text and results.[5]

3.2 Pricing Forward Contracts and Managing Counterparty Risk

Recall that Section 2.3 introduced a simple example of a forward contract on a stock. How can the fair price F of such a forward contract maturing at time T be determined? 'Fair' in this context will mean that the contract has an initial value of 0. It might seem intuitive that the forward price is a function of the price distribution of the stock at time T. This, however, is not the case. Under the above stated assumptions the forward price can be derived as follows:

Theorem 3.1 (No-arbitrage price of a forward contract). *Let S_t be the price of a stock at time $t \in [0, T]$, and r be the risk-free interest rate. If the stock does not pay dividends up to time T, the no-arbitrage forward price $F(t, T)$ at time t is given by*

$$F(t, T) = S_t e^{r(T-t)}. \tag{3.1}$$

Proof. Assume that $F(t, T) > S_t e^{r(T-t)}$. We can implement the following arbitrage strategy producing a non-zero cash flow only at time T.

Position/time	t	T
Sell forward with maturity T	0	$F(t, T) - S_T$
Borrow cash S_t over $[t, T]$	S_t	$-S_t e^{r(T-t)}$
Buy stock at time t, sell it at time T	$-S_t$	S_T
Total cash flow of portfolio	0	$F(t, T) - S_t e^{r(T-t)} > 0$

Hence, $F(t, T) > S_t e^{r(T-t)}$ cannot hold under the no-arbitrage condition. Similarly, assuming $F(t, T) < S_t e^{r(T-t)}$ leads to the following arbitrage portfolio.

Position/time	t	T
Buy forward with maturity T	0	$S_T - F(t, T)$
Borrow and sell stock at time t, return it at time T	S_t	$-S_T$
Deposit cash S_t over $[t, T]$	$-S_t$	$S_t e^{r(T-t)}$
Total cash flow of portfolio	0	$S_t e^{r(T-t)} - F(t, T) > 0$

[5] In practice dividend payments are often modeled in such a way that the properties of the underlying model do not change much.

Thus, the forward price can only be (3.1).[6] In case of interest rates differing for various maturities, the above constant interest rate r can simply be replaced by $r(t, T)$ and the arguments still hold true.[7] □

Example

Assume a stock initially trades at $S_0 = 100$, and the borrowing/lending rate is $r = 0.05$. You take a long position in a forward contract to buy one stock at $F(0, 1) = 100 \cdot e^{0.05 \cdot 1} = 105.13$ in one year from now. After 6 months, the stock price surprisingly increases to $S_{0.5} = 200$. You can now take a short position in a new forward with maturity $T = 1$, and the forward price would be $F(0.5, 1) = 200 \cdot e^{0.05 \cdot 0.5} = 205.07$. At time 1, you now buy one stock at 105.13 and sell one at 205.07. Thus, you will make a profit $205.07 - 105.13 = 99.94$. The initial forward contract therefore has considerable value at time 0.5. However, there remains the risk that your counterparty in the first forward contract will not fulfil its financial obligations.

The risk of the counterparty not fulfilling a contract is referred to as *counterparty risk*, and is a form of credit risk. Recall that futures are standardized and exchange-traded forward contracts. Futures exchanges offer a mechanism to lower counterparty risk. When entering into a futures contract, both counterparties will be asked to open a *margin account* with a clearinghouse (e.g. LCH.Clearnet). Each party deposits a certain initial amount as *initial margin*. The future is then *marked-to-market* (i.e. the profit/loss ($P\&L$) at maturity T is calculated under the assumption of closing one's future position today) daily, and the margin accounts are adjusted according to the daily loss/profit made on the position. This mechanism will become clear from the example below. If the margin account balance drops below a certain *maintenance margin*, the clearinghouse will issue a *margin call* in which the future counterparty will be asked to deposit additional funds (the *variation margin*) into his/her account to re-reach the initial margin. If the counterparty is not able to do so, the future position will be closed by the clearinghouse. Funds in excess of the initial margin can typically be withdrawn. Note that margin accounts in principle limit the loss from counterparty risk to price moves of one day. The following example illustrates the functioning of a margin account.

Example (Margin account)

On July 5, you take a long position in a futures contract to buy 100 underlyings in 6 months at a $F(0, 0.5) = 600$. The clearinghouse sets the initial margin at 5,000 and the maintenance margin at 3,500. The following table shows the development of the account balance, based on the changes in the futures prices; the margin account (MA) balance on any day is stated before margin calls and withdrawals.

---

[6]In the presence of income from the underlying asset (e.g. dividends for a stock), storage costs or transportation costs, this formula will no longer hold (see Hull [41] for a discussion).

[7]Again under the assumption that the lending and borrowing rates are equal.

Date	F	Daily gain (loss)	MA Balance	Margin Call	Withdrawal
July 5	600		5,000		
July 6	597	(300)	4,700		
July 8	560	(3,700)	1,000	4,000	
July 9	565	500	5,500		(500)
July 10	565	0	5,000		

3.3 Bootstrapping

The following simple example illustrates the concept of *bootstrapping*.

Example (Zero rates from coupon bonds)

Let $P(t_0, T)$ be the price at time t_0 of a zero-coupon bond paying 1 at time T. If we interpret the price as a discount factor at (zero-)rate $r(t_0, T)$, we can simply write $P(t_0, T) = e^{-r(t_0,T)\cdot(T-t_0)}$. Hence, once the zero-coupon bond prices have been determined, it is straightforward to translate the prices into zero-rates. The issue is now that only short-term bonds (e.g. up to a maturity of 12 months) are usually structured without coupons. The following could be market data of US government T-Bills (maturities $\leq$ 12 months) and T-Notes (maturities $> 1 - 10$ years).

Maturity	Type	Coupon % p.a.	Coupon type	Price at t_0
6 months	no coupon	n/a	n/a	99.26c
12 months	no coupon	n/a	n/a	99.28c
1.5 years	coupon	1.4 %	semi-annual	99.14c
2 years	coupon	2 %	semi-annual	99.55c

Neglect day-count issues by assuming that coupons are paid at $t_0 + 0.5, t_0 + 1, t_0 + 1.5, t_0 + 2$. If we see the bond prices as the sum of zero-coupon bonds of different sizes, we can write the following linear system of equations for bonds of face value 100:

$$\begin{bmatrix} 100 & 0 & 0 & 0 \\ 0 & 100 & 0 & 0 \\ 0.7 & 0.7 & 100.7 & 0 \\ 1 & 1 & 1 & 101 \end{bmatrix} \cdot \begin{bmatrix} P(t_0, t_0 + 0.5) \\ P(t_0, t_0 + 1) \\ P(t_0, t_0 + 1.5) \\ P(t_0, t_0 + 2) \end{bmatrix} = \begin{bmatrix} 99.26 \\ 99.28 \\ 99.14 \\ 99.55 \end{bmatrix}.$$

Due to the structure of this system it is easy to see that $P(t_0, t_0+0.5) = 0.9926$ and $P(t_0, t_0+1) = 0.9928$. Substituting these prices in the third equation of the system we obtain that $P(t_0, t_0 + 1.5) = 0.9707$, and substituting all three prices into the fourth equation of the system finally leads to $P(t_0, t_0 + 2) = 0.9564$. This technique of starting with low maturities, and successively determining the implied zero-coupon bond prices for increasing maturities is called bootstrapping. With $r(t_0, T) = -\frac{\ln(P(t_0,T))}{T-t_0}$ it is straightforward to find $r(t_0, t_0 + 0.5) = 1.5\,\%$, $r(t_0, t_0 + 1) = 1.75\,\%$, $r(t_0, t_0 + 1.5) = 2\,\%$ and $r(t_0, t_0 + 0.5) = 2.25\,\%$, which indicates an *upward-sloping zero-curve*.

Let us now turn to more general considerations, using a plain vanilla Libor interest rate swap with notional 1 (cf. Section 2.4). What is the present value $PV_s(t_0)$ of the swap at the initial time t_0? Neglecting counterparty risk, the receiver (floating payer, fixed receiver) of the swap finds $PV_s = PV_{\text{fix}} - PV_{\text{fl}}$, with PV_{fix} and PV_{fl} being the present values of the fixed and floating leg at t_0, respectively. PV_{fix} can be obtained by discounting with the interest curve,

$$PV_{\text{fix}} = \sum_{i=1}^{n} e^{-r(t_0,t_i)\cdot(t_i-t_0)} \cdot \text{sr},$$

where sr is the *swap rate*, t_i $(i = 1,\ldots,n)$ are the payment times and n is the number of coupon payments. For the determination of PV_{fl}, note that the floating leg cash flows are the same as from a plain vanilla floating bond (cf. Section 1.4), with the exception of the final principal repayment of 1. 'Artificially' including the payment of the notional amount 1 in both legs does obviously not change PV_s. We indicate the payment of the notional in the present values by a '+' and write

$$PV_{\text{fix}}^{+} = \sum_{i=1}^{n} e^{-r(t_0,t_i)\cdot(t_i-t_0)} \cdot \text{sr} + e^{-r(t_0,t_n)\cdot(t_n-t_0)} \quad \text{and} \quad PV_{\text{fl}}^{+} = 1 \tag{3.2}$$

Recall from Chapter 1 that the fair price of a vanilla floater equals its principal amount initially and on coupon payment dates. It then follows that

$$PV_s = PV_s^{+} = \sum_{i=1}^{n} e^{-r(t_0,t_i)\cdot(t_i-t_0)} \cdot \text{sr} + e^{-r(t_0,t_n)\cdot(t_n-t_0)} - 1.$$

As the swap rate sr is chosen such that $PV_s = 0$, setting the above to zero and solving with respect to sr produces the swap rate as a function of the prices of zero-coupon bonds. The value of the swap at later payment dates can be obtained similarly to (3.2). One simply replaces t_0 by t_k and starts the summation with $i = k$.

Finally, how is the value determined for times in between payment dates?[8] Let t be a valuation date with $t_{k-1} < t < t_k$. It then holds that

$$PV_{\text{fix}}(t) = \sum_{i=k}^{n} e^{-r(t,t_i)\cdot(t_i-t)} \cdot \text{sr}.$$

If the notional was also exchanged, the present value of the floating leg at time t_k would equal its notional, so that we find

$$PV_{\text{fl}}(t) = e^{-r(t,t_k)\cdot(t_k-t)} \cdot (\text{Libor}_{t_{k-1}} + 1) - e^{-r(t,t_n)\cdot(t_n-t)},$$

with Libor_t denoting the appropriate Libor rate fixed at time t.

[8]For example, if a company has to report its assets and liabilities in between payment dates, it will also have to report the value of its swaps.

In practice swap rates will be available for a large range of maturities and the zero curve can be computed from the quoted swap rates.[9]

Bootstrapping Method for Swap Rates

Assume that we know the swap rates sr_i for a term of $i = 1, 2, \ldots$ years at time t_0, and we use here $t_i = t_0 + i$. The zero rates $r(t_0, t_i)$ can then be extracted from the data points inductively:

1. $k = 1$: $PV_{\mathrm{fl}}^{+}(t_0) = 1$. Thus, one has to solve

$$1 = (1 + \mathrm{sr}_1) \cdot e^{-r(t_0,t_1)\cdot 1}.$$

If $\mathrm{sr}_1 \geq 0$, one obtains the unique solution $r(t_0, t_1) = -\ln\left(\frac{1}{1+\mathrm{sr}_1}\right) > 0$.

2. Induction step. Assume that the zero rates $r(t_0, t_i)$ have been computed for $i = 1, \ldots, k-1$. To obtain $r(t_0, t_k)$, the following equation has to be solved:

$$1 = e^{-r(t_0,t_k)\cdot(t_k - t_0)} \cdot (\mathrm{sr}_k + 1) + \sum_{i=1}^{k-1} e^{-r(t_0,t_i)\cdot(t_i - t_0)} \cdot \mathrm{sr}_k .$$

If the k-year swap rate sr_k is so large that the second term above is ≥ 1, then there exists no finite solution $r(t_0, t_k)$. However, this problem will not arise in liquid markets (see Exercise 3).

Again, the procedure of inductively determining zero-rates implied by the swap rates is referred to as bootstrapping. In practice, the construction of the entire zero curve will be based on a grid of reference points extracted from market prices as above. This grid will usually be narrower for shorter maturities. As swap rates are often only quoted for maturities in excess of one year, Libor/Euribor rates (such as ON (overnight), 1D, 7D, 1M, 2M, 3M, 6M, 9M, 12M), or connected futures, can be used to construct the short end of the zero curve.

3.4 Forward Rate Agreements (FRAs)

Forward contracts are not only written on currencies, stocks or stock indices, but also on interest rates. The simplest such contract is a *forward rate agreement* (FRA). At time t_0, the FRA defines a time interval $[y, z]$, $t_0 < y < z$, over which some

[9]Swap rates are quoted for a larger range of maturities than zero-coupon bonds (*quoted* in this context means that information providers, such as Reuters or Bloomberg, continuously publish current prices at which market makers, such as large banks, offer the respective product). Note, however, that swap rates will not be quoted for all maturities if one considers very long terms, e.g. there will be no liquid trading in 34-year swaps. In such a case one can interpolate the lacking rates from other data points.

reference interest rate (e.g. Libor or Euribor) is exchanged for a fixed rate. Thus, a plain vanilla interest rate swap, which repeatedly exchanges fixed for floating interest payments, could be seen as a portfolio of FRAs. An FRA contract will mostly be cash-settled at its *effective date* y, as the reference rate will already be known by then. In particular, a $y \times z$ FRA that is settled at its effective date y will pay (here: to the fixed payer)

$$N \cdot \frac{(r_{\mathrm{ref}} - r_{\mathrm{fixed}}) \cdot \mathrm{DCF}}{1 + r_{\mathrm{ref}} \cdot \mathrm{DCF}},$$

where N is the notional amount, r_{ref} is the reference interest rate for the period $[y, z]$ at time y, r_{fixed} is the fixed rate as agreed in the FRA and DCF is the day-count fraction of the period $[y, z]$ (see Section 1.2).

Example

To settle a '3 × 6 FRA', one would exchange the Libor3M rate in 3 months from now for the fixed interest rate on the notional amount for the period starting in 3 months and ending in 6 months. As interest payments are usually made at the end of the period, the settlement amount has to be discounted if paid out already after 3 months (cf. the above formula). A '9 × 15 FRA' will use the Libor6M (15M − 9M = 6M) in 9 months from now as reference rate, the effective date would be 9 months from now, and the *termination date* 15 months. Note that in order to ensure liquidity in the markets, the British Bankers Association[10] offers standard definitions and documentation for FRA contracts.

3.5 Key Takeaways, References and Exercises

Key Takeaways

After working through this chapter you should understand and be able to explain the following terms and concepts.

- Financial arbitrage, cross-market arbitrage
- Common assumptions when modelling financial markets (6 were listed)
- The no-arbitrage price of forwards, margin account (initial margin, maintenance margin, variation margin)
- Bootstrapping for (a) ZCB/coupon bonds (as in the example) and (b) swap rates
- FRAs, the pay-off of an FRA

[10] www.bba.org.uk

References

A rigorous introduction to the notion of no-arbitrage is given in Delbaen & Schachermayer [20]. Hull [41] explains how to derive no-arbitrage prices for many types of derivatives.

Exercises

1. The exchange rate between GBP and EUR today shall be 1 GBP = 1.4 EUR, the 5-year interest rate (continuously compounded) shall be $r_{5;\mathrm{GBP}} = 5.6\,\%$ for GBP, and $r_{5;\mathrm{EUR}} = 5.2\,\%$ for EUR. Use no-arbitrage arguments to determine the fair price of a GBP/EUR FX-forward with a maturity of 5 years. What financial instruments will be required to construct the no-arbitrage portfolio?
2. The gold spot price today shall be 1,500 EUR per ounce and the forward price to purchase gold in one year from now shall be 1,700 EUR. Explain how one could generate risk-free profits (arbitrage), if money can be borrowed at 5 % p.a.? (Assume that there are no costs attached to storing gold – in practice gold is treated like a currency (code XAU), since its storage costs can be neglected relative to its traded price.)
3. Determine how to make arbitrage profit if the swap rates for maturities 1-year to 14-years are 3 % flat, and the 15-year swap rate is 10 %. Generalize this simple example, and show that one will always be able to find an arbitrage opportunity if the summation term in the induction step of the bootstrapping method on page 33 is greater than 1. (Assume that zero-coupon bonds and vanilla floaters are liquid for maturities of 1 year to 15 years.)
4. Assume that swaps are traded for 1-, 2- and 3-year maturities, and their market swap rates are all 4%. Determine the price of a bond with annual coupon payments of 5% and a maturity of 3 years (assume that the credit risk of the bond corresponds to the one assumed for producing the swap rates). How could you generate arbitrage profits if the traded price differed from the price you obtained? (Assume that vanilla floaters with a maturity of 3 years are liquid in the market.)
5. Assume that two zero-coupon bonds with different maturities T_1 and T_2 are liquid instruments with given prices. Determine the no-arbitrage interest rate $r(T_1, T_2)$ (also referred to as *forward rate*[11]) for continuous compounding.

Exercises with Mathematica and UnRisk

6. Apply the UnRisk commands `MakeSwapCurve` and `MakeYieldCurve` to generate swap curves. Use the attained swap curves to produce the corresponding zero curves through bootstrapping. Plot both curves. Use `Manipulate` to create a scroll bar which allows to move single nodes of the swap curve. What is the related effect on the zero curve?

[11]Forward rates are interest rates for periods that start in the future, and they can be obtained from current interest curves.

European and American Options

4

We discussed in Chapter 2 that an option gives the buyer a particular right which can lead to financial upsides in the future, without including any obligations. Hence, there must be a positive price for obtaining this right, and we will now aim to determine this price. Figure 4.1 shows market prices of European call options on the Euro-Stoxx50 index as a function of the strike for various expiries (the plot also includes the pay-off that would be attained assuming the stock price at expiry was S_0).

It is intuitive that call prices will be lower for higher strikes, Figure 4.1 also shows higher option prices for longer times until expiry. Although actual option prices are the result of demand and supply, financial mathematics can answer many structural questions. How sensitive is an option price to changes in price-driving parameters? What is the fair price of an illiquid option in the OTC market? To answer such questions one typically chooses a particular stochastic model to describe the dynamics of the price of the underlying. In addition, we will discuss in Section 4.1 that simple no-arbitrage considerations allow to derive model-independent identities and bounds for option prices. Specific model-dependent results will follow in later chapters. To facilitate notation, the considerations in Chapters 4–8 will focus on stock options.

It is obvious that the option price has to be determined in a way such that buyers and sellers agree on entering into the contract. However, if it is possible to construct a portfolio from cash and stocks, such that the value of that portfolio at time T equals the pay-off of the option, then today's value of the portfolio is the *fair price* of the European option. We will see in the following chapters that such a *replication* of the option pay-off through a portfolio of cash and stocks is possible under certain model assumptions. This will require the ability to continuously adjust the portfolio by buying and selling stocks.[1]

---

[1] A *static* portfolio, i.e. one that is chosen initially but then not adjusted anymore prior to expiry, can typically not replicate option pay-offs, as the option price is not a linear function of today's stock price. Note that this is different for forward contracts.

H. Albrecher et al., *Introduction to Quantitative Methods for Financial Markets*, Compact Textbooks in Mathematics, DOI 10.1007/978-3-0348-0519-3_4,

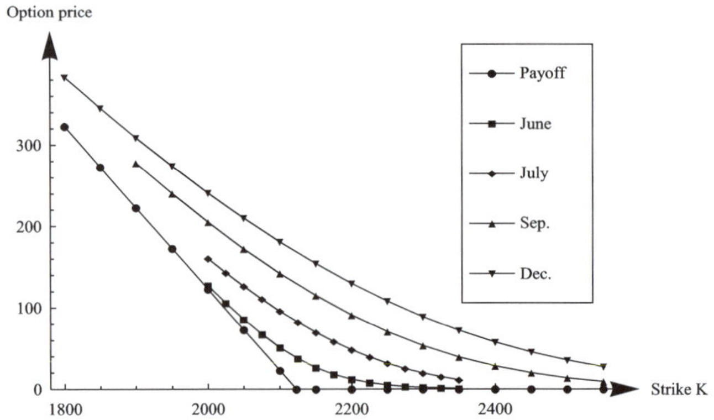

Fig. 4.1 Prices and pay-off of calls on the EStoxx50 index on 6 June 2012 ($S_0 = 2123$) with expiries at (end of) June, July, September and December 2012

The resulting trading strategy that replicates the option is a *hedge* against the risk that arises when selling the European option, i.e. the risk that S_T will be smaller/larger than K.

4.1 Put-Call Parity, Bounds for Option Prices

Let C_t (P_t) be the price of a European call option (put option) at time t. At expiry T, it follows from the definitions that

$$C_T - P_T = (S_T - K)^+ - (K - S_T)^+ = S_T - K.$$

Under the assumptions set out in Section 3.1, it can be shown that arbitrage opportunities can only be ruled out if the following holds at all times $t \leq T$:

Theorem 4.1 (Put-Call Parity). *For plain vanilla European options with expiry T and strike K, and some constant risk-free interest rate r, it must hold that*

$$C_t - P_t = S_t - Ke^{-r(T-t)} \qquad \text{for all } t \in [0, T]. \tag{4.1}$$

Proof. Consider the following two portfolios at time t:

Portfolio A: one European call option and $Ke^{-r(T-t)}$ of cash

Portfolio B: one European put option and one stock

The following table shows the value of the portfolios A and B at time T, depending on S_T being smaller than K or not.

portfolio/case	$S_T \geq K$	$S_T < K$
A	$(S_T - K) + K$	$0 + K$
B	$0 + S_T$	$(K - S_T) + S_T$

As both portfolios produce the same value at time T for any outcome of S_T, it follows from no-arbitrage arguments that their values must be equal also at time t, i.e. for all $0 \leq t \leq T$,

$$C_t + Ke^{-r(T-t)} = P_t + S_t. \qquad \square$$

For stocks that do not pay dividends prior to expiry of the considered options, the price of a European put option can be determined by the put-call parity if the price of the otherwise identical European call option is known. In the following we can thus restrict our analysis to call options.

Trivial bounds for the option prices (see Exercise 1),

$$0 \leq C_t \leq S_t \quad \text{and} \quad 0 \leq P_t \leq K\,e^{-r(T-t)}, \tag{4.2}$$

in combination with the put-call parity, imply the following.

Conclusion

In a no-arbitrage market with no dividend payments, it holds that

$$\max\left(S_t - Ke^{-r(T-t)}, 0\right) \leq C_t \leq S_t.$$

and

$$\max\left(Ke^{-r(T-t)} - S_t, 0\right) \leq P_t \leq K\,e^{-r(T-t)}.$$

These bounds are not particularly sharp, but they are solely based on the fundamental assumptions in Section 3.1 and are hence independent of the choice of market model.

Up to now, it has been assumed that stocks do not pay dividends to stock holders. As the time to expiry of options often does not exceed 12 months, one might be able to project dividend payments prior to expiry with satisfactory accuracy. Hence, let D_t be the present value at time t of the dividend payments up to expiry T. The put-call parity then needs to be adapted to (see Exercise 3)

$$C_t - P_t = S_t - D_t - Ke^{-r(T-t)} \qquad \text{for all} \quad t \in [0, T] \tag{4.3}$$

and one thus obtains the bounds

$$C_t \geq S_t - D_t - Ke^{-r(T-t)} \qquad \text{for all } t \in [0, T] \tag{4.4}$$

and

$$P_t \geq D_t + Ke^{-r(T-t)} - S_t \qquad \text{for all } t \in [0, T]. \tag{4.5}$$

4.2 Some Option Trading Strategies

Section 2.5 discussed the pay-off of single European call and put options depending on the stock price of the underlying at expiry T. In order to derive a profit/loss curve, the pay-offs of the option portfolios have to be adjusted for the initial price of the respective options. We now consider combinations of long and short positions to tailor pay-off and profit/loss ('P&L') profiles as functions of the stock price at expiry T.

Example

- A *bull spread* is a portfolio containing a long position in a call with strike K_1 and a short position in a call with the same expiry and strike $K_2 > K_1$. An investor using a bull spread expects a high stock price at time T (see Figure 4.2). Note that the pay-off function can also be produced by using two put options (see Exercise 6).
- Taking a short position in a call with strike K_1 and a long position in a call option with strike K_2 (again: $K_2 > K_1$) gives a *bear spread*[2]. This kind of spread will produce a profit for low stock prices at time T (see Figure 4.2). Note that for both bull and bear spreads, potentially high profits are given up for limiting the strategy's downside.
- A *butterfly spread* consists of four options with altogether three different strikes. The portfolio contains two long call options, one with lower strike K_1 and one with higher strike K_3, and two short call options with strike $K_2 = (K_1 + K_3)/2$. Typically K_2 will be close to the initial stock price S_0. The resulting P&L curve as a function of S_T is shown in Figure 4.3. A butterfly spread produces a profit if the stock price S_T at expiry is close to K_2, and a relatively small loss otherwise. Such a strategy can be implemented by investors who expect only small changes in the underlying stock price. Note that a butterfly spread with identical P&L profile can also be constructed by using four put options (see Figure 4.3). It follows from the no-arbitrage principle that the initial cost of compiling the portfolio from put or call options must be the same.

The above example suggests that one can arbitrarily combine positions in call and put options to produce (almost) any desired pay-off profiles at time T.[3] For example, the location and the height of the spike in a butterfly spread can then be adjusted by varying the number of options used and their strikes. The set of possible pay-off profiles can be even further extended when combining options with different expiry dates.

---

[2]A *bullish (bearish) market* is a financial market in which prices are expected to rise (fall). The 'bulls' are current buyers driving prices up higher through their additional demand, while 'bears' sell positions, which results in dropping prices. The terms 'Hausse' ('Baisse') market are also used sometimes.

[3]On a practical note, put and call options might not be liquid in the market for some strikes.

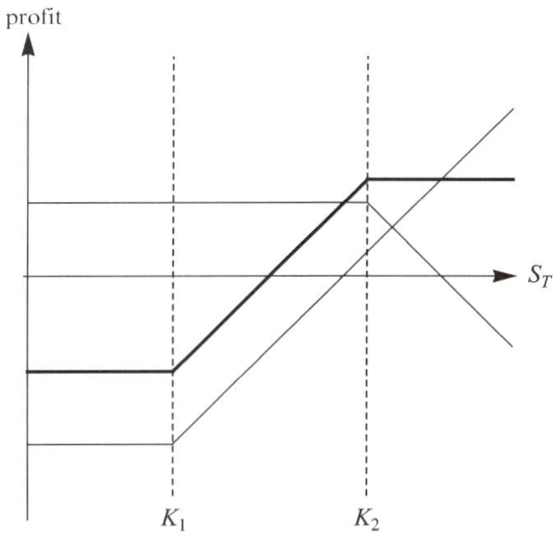

Fig. 4.2 Profit/loss profile of a bull-spread (left) and a bear-spread (right)

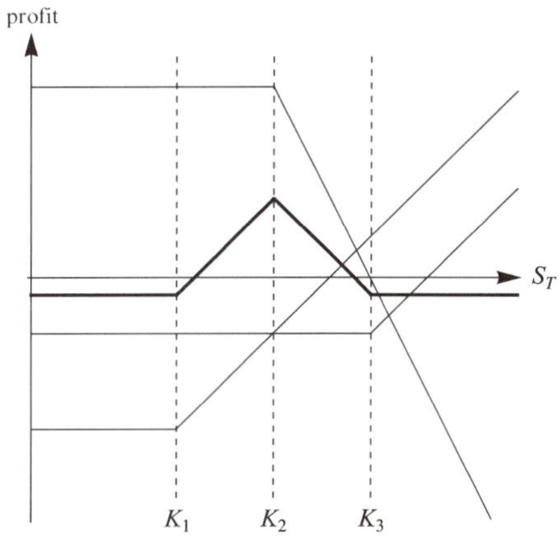

Fig. 4.3 Profit/loss (P&L) profile of a butterfly-spread with calls (left) and puts (right)

4.3 American Options

In contrast to European options, American options can be exercised at any point in time up to expiry T. This additional feature requires to first determine the optimal exercise time when pricing the option, which makes price computations more cumbersome. For this reason, in this introductory text we will only deal with some general properties of American options.

In a no-arbitrage market the value of an American option must be at least the value of the corresponding European option, since the American option provides greater flexibility with respect to the timing of the exercise.

Theorem 4.2. *Let $C_0(E)$, $P_0(E)$ and $C_0(A)$, $P_0(A)$ be the prices of European (E) and American (A) call and put options with strike price K and expiry date T. In a no-arbitrage market, the following inequalities must hold:*

$$0 \leq C_0(E) \leq C_0(A) \leq S_0,$$
$$0 \leq P_0(E) \leq P_0(A) \leq K,$$
$$P_0(A) \geq \max(K - S_0, 0).$$

Proof. Assume that $C_0(E) > C_0(A)$. One could then generate risk-free profit by selling a European call option and buying an American call option, and pocketing the price difference $C_0(E)-C_0(A) > 0$. The American option can then be held until time T, where it will have the same pay-off as the European option, so that arbitrage profit has been made. If $C_0(A) > S_0$, one can simply buy a stock at S_0 and sell a call option. The initial profit is now risk-free, as the potential obligations under the option contract are fully covered by the long position in the stock. The second chain of inequalities for put options can be justified analogously. Finally, the price of an American put option must be positive and at least its pay-off for immediate exercise, $K - S_0$, as one can otherwise again find an arbitrage strategy. □

The attained price bounds are relatively wide, however, note that they are again model-independent. Using these bounds allows to find surprisingly simple relationships between American and European call prices, such as the following.

Theorem 4.3. *If the underlying stock does not pay dividends up to expiry of the options, then for a no-arbitrage market with $r > 0$ we have*

$$C_0(E) = C_0(A). \tag{4.6}$$

Proof. Due to $r > 0$ and the put-call parity for European options, we find $C_0(A) \geq C_0(E) \geq S_0 - Ke^{-rT} > S_0 - K$. The initial price of the American call must therefore be at least its pay-off upon immediate exercise. Thus, it is not optimal to immediately exercise the option as long as the option holder prefers 'more' over 'less'. Subsequently, a similar argument can be applied for all times $t < T$ for which $C_t(E) \geq S_t - Ke^{-r(T-t)}$ and, hence, $C_t(A) > S_t - K$. It is therefore never optimal to exercise the American call prior to its expiry date. At expiry, the American and European option have the same pay-off, and (4.6) follows from the no-arbitrage condition. □

An American call option on a stock that does not pay dividends should therefore not be exercised before maturity. If the option buyer adopts at some point before expiry the view that the stock price was extraordinarily high and would hence like to exercise the option, it will be more profitable to simply sell the option in the market than to use the exercise right. Early exercise of an American put option on a stock that does not pay dividends can, on the other hand, be optimal. It therefore follows for $r > 0$ that $P_0(A) > P_0(E)$.

For American options, there is no put-call parity. However, one can establish the following relationships (see Exercise 13):

$$S_t - K < C_t(A) - P_t(A) < S_t - Ke^{-r(T-t)} \qquad \text{for all } t \in [0, T]. \tag{4.7}$$

Whenever dividend payments are expected, it can be of advantage to also exercise call options early. This can particularly be the case prior to a dividend payment which would typically be followed by a drop in the stock price.

Instead of (4.7), one finds for known present value D_t of the dividend payments that

$$S_t - D_t - K < C_t(A) - P_t(A) < S_t - Ke^{-r(T-t)} \qquad \text{for all } t \in [0, T]. \tag{4.8}$$

4.4 Key Takeaways, References and Exercises

Key Takeaways

After working through this chapter you should understand and be able to explain the following terms and concepts:

- European vs. American options
- The Put-Call Parity
- Model-independent bounds to option prices as implied by the Put-Call Parity
- Option portfolio strategies, including bull, bear and butterfly spreads
- The relationship between $C_0(E)$ and $C_0(A)$ in the absence of dividend payments
- Early exercise of American (a) call and (b) put options

References

A detailed overview of European and American options, as well as option-based trading strategies, can be found in many sources, including Wilmott [75] or Capinski & Zastawniak [14].

Exercises

1. Prove that the no-arbitrage bounds (4.2) hold.
2. Provide a graphical proof of the put-call parity for European options at expiry T.
3. In the presence of dividend payments, prove the put-call parity relation (4.3) and the bounds to the option prices (4.4) and (4.5).

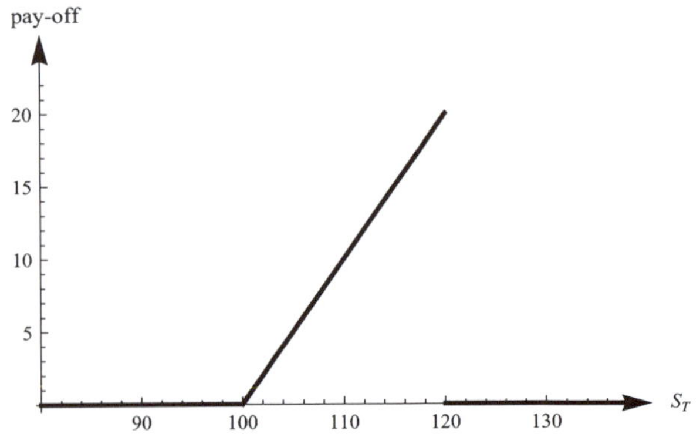

Fig. 4.4 Pay-off of the option of Exercise 12

4. Investigate what profit/loss profiles can be attained as function of S_T, if a portfolio consists of a stock and a long (or short) call (or put) option.
5. (a) Determine a lower bound of the price of a call option on a stock that does not pay dividends. Assume the option parameters $T = 4$ months, $K = 25$ EUR and a risk-free interest rate $r = 5\,\%$ p.a. (continuous compounding). The initial stock price shall be $S_0 = 28$ EUR.
   (b) Let the price of an American call option on a stock that does not pay dividends be 4 EUR. The initial stock price shall be $S_0 = 31$ EUR, the strike $K = 30$ EUR, the expiry $T = 3$ months, and the risk-free interest rate $r = 5\,\%$ p.a. (continuous compounding). Determine lower and upper bounds for an American put option with identical parameters.
6. Replicate the pay-off of a bull and a bear spread, respectively, by using only put options. Does the initial net cash flow differ compared to a construction by call options?
7. Let C_1, C_2 and C_3 be the prices of European call options with strikes K_1, K_2 and K_3, respectively. Assume $K_3 > K_2 > K_1$ and $K_3 - K_2 = K_2 - K_1$. Prove the convexity property

$$C_2 \leq 0.5(C_1 + C_3),$$

given that all options have the same expiry T.
8. Three European put options on a stock have the same expiry date T, and strikes 55 EUR, 60 EUR and 65 EUR, respectively. The options price at 3 EUR, 5 EUR and 8 EUR in the market. How can you use the options to construct a butterfly spread? Produce a table defining the profit/loss profile of such a trading strategy. For what values S_T does the butterfly spread lead to a loss?
9. (a) Use the put-call parity to show that the initial costs of assembling a butterfly spread based on only European put options or only European call options are equal.
   (b) How can a forward contract on a stock with forward price F and maturity T be replicated by the use of options?
10. The price of a European call option with expiry $T = 6$ months and strike $K = 30$ EUR shall be 2 EUR. Let $S_0 = 29$ EUR, and let the risk-free interest rate be $r = 0.08$. Determine the price of a European put option with expiry $T = 6$ months and strike $K = 30$ EUR, if a dividend payment of 0.5 EUR will be made in 2 and in 5 months.
11. A market participant might have the view that the price of a stock will change drastically from its price today S_0, either upwards or downwards. This might be due to the uncertain outcome of a binary event, like the winning or losing of a court case. Based on this view, the market participant goes long on a put with strike $S_0 - \Delta$ and long on a call with strike $S_0 - \Delta$, both with expiry T. Such an option portfolio is called *strangle*. Draw the pay-off profile at time T of this portfolio. For each option in the portfolio, does the counterparty face unbounded

potential downside? How do you have to choose Δ, if you would like to have a pay-off of 10 from the portfolio in case of $S_T = 80$, given that $S_0 = 100$?

12. Consider an option with pay-off

$$\text{pay-off} = (S_T - 100)^+ \, 1_{\{S_T \leq 120\}},$$

where 1 is the indicator function (see Figure 4.4). Note that this is a simplified version of a barrier option. Assume that European options with the same expiry are liquid for all strikes. Find a portfolio consisting of European options, whose pay-off does not differ from the pay-off of the option outside the interval [119.5, 120.5].

13. Prove the inequalities (4.7) and (4.8) by using no-arbitrage arguments.

The Binomial Option Pricing Model

5

The following chapters will be dedicated to the stochastic modeling of price movements of financial assets. Chapters 5 to 8 will focus on stocks[1], while Chapter 9 will deal with interest rates.

5.1 A One-Period Option Pricing Model

To introduce models describing the price movements of stocks, we will start with a simple market model with only one trading period, initial time 0 and time horizon T. The stock price S_T at the end of this period is modeled by the random variable

$$S_T = \begin{cases} s_1 & \text{with probability } p, \\ s_2 & \text{with probability } 1 - p. \end{cases}$$

Consider a European option with pay-off $V_T(S_T = s_1) = v_1$ and $V_T(S_T = s_2) = v_2$. The idea is now to construct a portfolio at time $t = 0$ that replicates the pay-off of the option at time $t = T$ and that consists of only the underlying stock and a risk-free account (e.g. cash or risk-free bonds) which earns interest at some fixed rate $r > 0$. If this portfolio contains θ_1 units of the stock and θ_0 monetary units of the risk-free account[2], then the value of the portfolio at time $t = 0$ is given by

$$V_0 = \theta_0 + \theta_1 S_0. \tag{5.1}$$

Replicating the pay-off of the option at $t = T$ by the portfolio (θ_0, θ_1) leads to the conditions

[1]Similar models are also often applied for exchange rates. This will not be further mentioned, but will be illustrated in some exercises.

[2]$\theta_0 \geq 0$ indicates a deposit in the cash account, while $\theta_0 < 0$ is a borrowing. Interest earned/paid is assumed at a constant rate $r > 0$.

H. Albrecher et al., *Introduction to Quantitative Methods for Financial Markets*, Compact Textbooks in Mathematics, DOI 10.1007/978-3-0348-0519-3_5,

$$v_1 = \theta_0 e^{rT} + \theta_1\, s_1,$$
$$v_2 = \theta_0 e^{rT} + \theta_1\, s_2.$$

This system of linear equations can easily be solved with respect to θ_0 and θ_1, and we find

$$\theta_1 = \frac{v_1 - v_2}{s_1 - s_2},$$
$$\theta_0 = e^{-rT}\left(v_1 - \frac{(v_1 - v_2)s_1}{s_1 - s_2}\right).$$

Based on (5.1), the value of the replicating portfolio at $t = 0$ is therefore

$$V_0 = S_0\left(\frac{v_1 - v_2}{s_1 - s_2}\right) + e^{-rT}\left(v_1 - \frac{(v_1 - v_2)s_1}{s_1 - s_2}\right). \tag{5.2}$$

V_0 must represent the fair value of the option at $t = 0$, as market participants could otherwise profit from arbitrage (cf. Exercise 1). Note that the price of the option is independent of the distribution of S_T (or, equivalently, of the probability p). This will be further discussed in the next section.

We have seen that it is possible to replicate the pay-off of the option by a portfolio consisting of θ_0 monetary units in a risk-free account and θ_1 stocks. This gives a strategy for hedging the option contract, as illustrated in the following example.

Example

The current price of a stock is $S_0 = 100$ EUR, and we have $T = 1$ year, $r = 0.05$, $s_1 = 130$ EUR and $s_2 = 80$ EUR. Consider a European call option on this stock with time-T pay-off $(S_T - K)^+$ and strike price $K = 110$ EUR. This implies option pay-offs $v_1 = 20$ EUR and $v_2 = 0$ EUR, leading to $\theta_1 = 0.4$ and $\theta_0 = -30.439$, so that we finally arrive at $V_0 = 9.561$ EUR for the initial price of the call option as per (5.2). The hedging strategy at time $t = 0$ for a short position in a call option is given as follows: the option seller receives the premium of 9.561 EUR and additionally borrows 30.439 EUR. The total amount of 40 EUR is then used to buy 0.4 units of the stock. At time $t = T$, there are two possible outcomes:

(i) $S_T = 130$. The call option is exercised. The option seller purchases some additional 0.6 units of the stock in the market at $0.6 \times 130 = 78$ EUR, and sells his holding of now one stock to the option buyer at 110 EUR. The excess of $110 - 78 = 32$ is used to repay the loan at $30.439\, e^{0.05} = 32$ EUR, so that the option seller is left with a net cash flow of 0 at time $t = T$.

(ii) $S_T = 80$. The option is not exercised. The 0.4 units of the stock are sold at $0.4 \times 80 = 32$ EUR, and the proceeds are used to repay the loan in full. The net payment of the option seller at time $t = T$ is again 0.

Therefore, the payments at time T from shorting the option and from holding the hedge portfolio cancel each other, if and only if the price of the option at $t = 0$ is set at $V_0 = 9.561$. Any other initial option price implies arbitrage opportunities (cf. Exercise 1).

Remark 5.1. The calculation of the no-arbitrage price above was straightforward since S_T could only take two possible values. If the set of possible values is extended to three or more, one can no longer determine a hedge portfolio (over $[0, T]$) consisting of stocks and the risk-free account.

5.2 The Principle of Risk-Neutral Valuation

As previously mentioned, the probabilities p and $1 - p$ of an increase or decrease of the stock price do not appear in formula (5.2).[3] This contradicts the possible intuition that the value of the call option would increase with larger values of p. In particular, the price of the option is *not* given by discounting the expected pay-off of the option. We now define the parameter

$$q = \frac{S_0 e^{rT} - s_2}{s_1 - s_2} \tag{5.3}$$

and introduce the random variable[4]

$$\tilde{S}_T = \begin{cases} s_1 & \text{with probability } q, \\ s_2 & \text{with probability } 1 - q. \end{cases}$$

It then follows from (5.2) that

$$V_0 = e^{-rT}\Big(qv_1 + (1-q)v_2\Big) = e^{-rT}\mathbb{E}[V_T(\tilde{S}_T)]. \tag{5.4}$$

Note that $\tilde{S}_T$ and S_T can only take the values s_1, s_2, but the probabilities of producing a particular realization are different. The distribution of $\tilde{S}_T$, which is obtained by modifying the distribution of S_T, is often referred to as *risk-neutral probability measure* $\mathbb{Q}$.[5] Hence, (5.4) can also be written as

$$V_0 = e^{-rT}\mathbb{E}^{\mathbb{Q}}[V_T]. \tag{5.5}$$

Using (5.3), it is straightforward to verify that

$$\mathbb{E}^{\mathbb{Q}}(S_T) = qs_1 + (1-q)s_2 = S_0 e^{rT}. \tag{5.6}$$

[3]The only required assumption here was that $p \in (0, 1)$.

[4]It can be easily shown through no-arbitrage arguments that $q \in (0, 1)$, so that $\tilde{S}_T$ indeed satisfies the definition of a random variable.

[5]The original distribution with probability p for a price increase is referred to as *physical probability measure* $\mathbb{P}$.

Thus, the stock price under the measure $\mathbb{Q}$ grows on average by the risk-free interest rate r, which explains the term 'risk-free measure'. Using $\mathbb{Q}$ therefore transforms the market into a 'fair game'.[6]

The above result is a particular case of the more general concept of *risk-neutral valuation*. The *Fundamental Theorem of Asset Pricing* states that in every discrete market with finite time horizon, which satisfies certain assumptions as listed in Section 3.1 (in particular, the no-arbitrage condition), there exists a risk-neutral measure $\mathbb{Q}$ (i.e. a re-weighting of the probabilities of the possible realizations), such that the prices of all derivatives can be calculated as their discounted expected pay-offs under the measure $\mathbb{Q}$.[7]

5.3 The Cox-Ross-Rubinstein Model

We will now turn to a generalization of the above one-period pricing framework. Let N be the number of equidistant trading times nT/N on the interval $[0, T]$ ($n = 1, \dots, N$), such that the price of a stock at these points in time is given by the distribution

$$S_{nT/N} = \begin{cases} (1+b)S_{(n-1)T/N} & \text{with probability } p, \\ (1+a)S_{(n-1)T/N} & \text{with probability } 1-p. \end{cases} \tag{5.7}$$

If r is the risk-free interest rate, no-arbitrage arguments lead to the condition $a < e^{rT/N} - 1 < b$ (cf. Exercise 2). This binomial type of model is often referred to as the *Cox-Ross-Rubinstein model* (short: *CRR model*, see Figure 5.1). The CRR set-up will now be applied to the pricing of derivatives that can be exercised at some given maturity T. Note that each of the N nodes at time $(N-1)T/N$ can be interpreted as a starting point of a one-period model (cf. Section 5.1).

This implies that the value of the derivative in each of these nodes is given as the discounted expected pay-off under the risk-neutral measure. Due to the symmetric structure (5.7) of the tree, the same risk-neutral probability

$$q = \frac{e^{rT/N} - 1 - a}{b - a} \tag{5.8}$$

is applied in each node. Note that the stock price cancels out in the above formula for q. After determining the price of the derivative in each node at time $(N-1)T/N$,

---

[6] S_T is also called *discounted martingale* under $\mathbb{Q}$, and $\mathbb{Q}$ is a *martingale measure*.

[7] For continuous-time models, the theorem still holds under certain restrictions. However, in such cases the risk-neutral measure will often not be unique and the selection of an appropriate measure will typically require additional assumptions (see Section 8.3). In practice, one sometimes starts directly with a risk-neutral model and calibrates that model to market data, which pre-selects a risk-neutral measure.

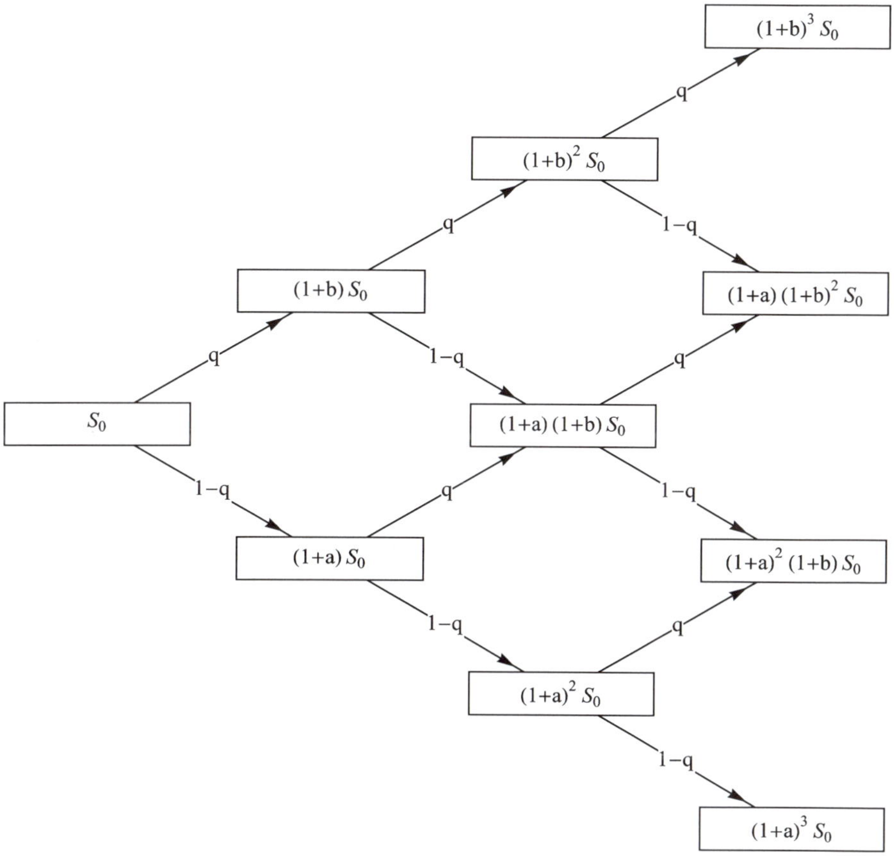

Fig. 5.1 Tree of possible outcomes in a CRR model for $N = 3$

one can repeat this procedure for the prices at time $t = (N-2)T/N$. Continuing to iteratively move backwards through the tree allows to ultimately find the price of the derivative at $t = 0$.

Consider first the case $N = 2$. The stock price in $t = T$ can then take one of the three values, $(1+b)^2 S_0$, $(1+a)(1+b)S_0$, or $(1+a)^2 S_0$. The corresponding pay-offs of the derivative shall be v_{22}, v_{21} and v_{11}. This yields

$$\begin{aligned} V_0 &= e^{-rT/2}\Big[q\left(e^{-rT/2}\Big(qv_{22} + (1-q)v_{21}\Big)\right) \\ &\quad + (1-q)\left(e^{-rT/2}\Big(qv_{21} + (1-q)v_{11}\Big)\right)\Big] \\ &= e^{-rT}\Big[q^2 v_{22} + 2q(1-q)v_{21} + (1-q)^2 v_{11}\Big]. \end{aligned}$$

One realizes that, due to the recursive structure of the CRR model, the value of a derivative can be written as the discounted expected pay-off under a binomial distribution with parameters N and q. For a European call with maturity T and strike price K, it follows for N time steps that

$$\begin{aligned} V_0 &= e^{-rT} \sum_{n=0}^{N} \binom{N}{n} q^n (1-q)^{N-n} \Big((1+b)^n (1+a)^{N-n} S_0 - K \Big)^+ \\ &= \sum_{n=m}^{N} \binom{N}{n} \frac{q^n}{e^{rnT/N}} \frac{(1-q)^{N-n}}{e^{r(N-n)T/N}} \Big((1+b)^n (1+a)^{N-n} S_0 \Big) \\ &\quad - K e^{-rT} \sum_{n=m}^{N} \binom{N}{n} q^n (1-q)^{N-n}, \end{aligned} \tag{5.9}$$

with m being the smallest positive integer that satisfies $S_0(1+b)^m(1+a)^{N-m} > K$.

Applying (5.8) and letting $q' = q(1+b)e^{-rT/N}$ yields $q' \in (0,1)$ and $1-q' = (1-q)(1+a)e^{-rT/N}$, so that the price of the European call option in this binomial model is

$$V_0 = S_0 \Psi(m; N, q') - K e^{-rT} \Psi(m; N, q), \tag{5.10}$$

with

$$\Psi(m; N, p) = \sum_{j=m}^{N} \binom{N}{j} p^j (1-p)^{N-j}.$$

Formula (5.10) is called the *Cox-Ross-Rubinstein formula* (or: binomial option pricing formula) for a European call option.

To summarize, the recursive application of the one-period model led to finding the initial price V_0 of a European call option (more generally, of a European-style derivative) in the binomial model. It is clear that the value $V_{nT/N}$ of the option at time $t = nT/N$ (at which time $S_{nT/N}$ is known) is given by

$$V_{nT/N} = S_{nT/N} \Psi(m_n; N-n, q') - K e^{-rT(1-n/N)} \Psi(m_n; N-n, q), \tag{5.11}$$

where m_n is the smallest positive integer that fulfills $S_{nT/N}(1+b)^{m_n}(1+a)^{N-n-m_n} > K$. The above also gives us a recipe for hedging the derivative for time $t = nT/N$. The portfolio $(\theta_0^{n-1}, \theta_1^{n-1})$ is held over the time period $[(n-1)T/N, nT/N)$, $n = 1, \ldots, N$, and it must replicate $V_{nT/N}$, which implies the condition

$$\theta_1^{n-1} S_{nT/N} + \theta_0^{n-1} e^{rT/N} = V_{nT/N}(S_{nT/N}).$$

Setting $S_{nT/N}$ to one of the two possible values $(1+a)S_{(n-1)T/N}$, $(1+b)S_{(n-1)T/N}$ and replacing $V_{nT/N}$ by the corresponding option value leads to unique portfolio weights $(\theta_0^{n-1}, \theta_1^{n-1})$. Generally, we find for $1 \le n \le N$ that

$$\theta_1^{n-1} = \frac{V_{nT/N}\left((1+b)S_{(n-1)T/N}\right) - V_{nT/N}\left((1+a)S_{(n-1)T/N}\right)}{S_{(n-1)T/N}(b-a)},$$

$$\theta_0^{n-1} = e^{-rT/N}\Bigg(V_{nT/N}\left((1+b)S_{(n-1)T/N}\right) - (1+b)\,\frac{V_{nT/N}\left((1+b)S_{(n-1)T/N}\right) - V_{nT/N}\left((1+a)S_{(n-1)T/N}\right)}{b-a}\Bigg).$$

We conclude that the hedging portfolio on $[(n-1)T/N, nT/N)$ consists of θ_0^{n-1} monetary units of the risk-free account and θ_1^{n-1} stocks. At time nT/N the portfolio is rebalanced to (θ_0^n, θ_1^n). Observe that no additional funds are required for the re-balancing. Such strategies are called *self-financing* and play a fundamental role in financial mathematics. Again, the physical probability measure influences neither the no-arbitrage price of the derivative, nor the weights of the hedging portfolio.

Remark 5.2. Despite its relatively simple structure, the Cox-Ross-Rubinstein model can be seen as a discretization of the well-known Black-Scholes model for an appropriate choice of parameters a and b (see Chapter 7). This is also the reason why binomial models are often used in practice for the development of numerical approximation algorithms, where obtaining explicit solutions in the corresponding Black-Scholes model proves challenging (cf. Exercise 7 and Section 6.3).[8]

5.4 Key Takeaways, References and Exercises

Key Takeaways

After working through this chapter you should understand and be able to explain the following terms and concepts:

- The no-arbitrage price of a European call in a one-period model with binary stock price $S_T \in \{s_1, s_2\}$
- Hedging in a one-period model
- The risk-neutral probability measure $\mathbb{Q}$, the price of a derivative as discounted expected pay-off under $\mathbb{Q}$
- The Cox-Ross-Rubinstein (CRR) model: pricing and hedging of European derivatives

[8] Such discretization techniques, however, require a good degree of caution (cf. Section 10.1).

References

Comprehensive discussions of discrete models can be found, e.g. in Shreve [71], Lamberton & Lapeyre [49], Föllmer & Schied [34] and Pascucci & Runggaldier [61]. These sources also discuss the martingale concept (see page 50), which can lead to an elegant description of trading strategies. This, however, is outside the scope of this introductory text. The interested reader is also referred to Delbaen & Schachermayer [20], where the Fundamental Theorems of Asset Pricing are derived in a mathematically rigorous way, and the connections between certain no-arbitrage conditions and the existence of a risk-neutral measure are analysed in detail.

Exercises

1. Find an arbitrage opportunity based on (5.2), if the market price of the option is not V_0.
2. Explain why arbitrage opportunities arise if the inequality $a < e^{rT/N} - 1 < b$ is violated.
3. The current price of a stock shall be 50 EUR and it is known that by the end of 2 months, the price will have moved to either 53 EUR or 48 EUR. The risk-free interest rate is 6 % p.a. (continuously compounded). Use no-arbitrage arguments to determine the price of a European call option with expiry $T = 2$ months and strike $K = 50$ EUR. What is the fair price of a European put option with the same parameters?
4. A stock trades at 100 EUR today. The risk-free interest rate shall be 5 % p.a. (continuously compounded) and the stock price shall be modeled in a CRR framework with parameters $a = -0.1$ and $b = 0.1$. Price changes shall occur quarterly. What are the fair prices of a European call and put option today, if the maturity is $T = 1$ year and the strike price is $K = 100$ EUR? Explain why the put-call parity is satisfied.
5. Compute the price V_0 of a European call option in the CRR model with parameters $T = 3$ days, $r = 0$, $K = 110$ EUR and $S_0 = 100$ EUR. Assume that the stock either rises or falls by 20 % on each trading day. Also provide a strategy to hedge the option. How can you generate a risk-free profit of 1 million EUR if the option is traded at $V_0 + 5$ EUR today?
6. The initial price S_0 of a stock shall be 25 EUR and it is known that the stock will price at $S_T = 23$ EUR or $S_T = 27$ EUR at time $T = 2$ months. The risk-free interest rate is 10 % p.a. (continuously compounded). What is the fair price of a derivative whose pay-off is $\log(S_T)$ at time T?
7. A stock trades at 100 EUR today and its price movements are described by a CRR model with monthly price jumps and parameters $b = 0.1$ and $a = -0.05$. Assume that the risk-free interest rate is $r = 0.05$. Use a backward recursion (as for a plain vanilla call) to determine the price of a barrier option that provides the same pay-off as a European call with strike price $K = 105$ EUR and maturity $T = 3$ months (the payment under the barrier option is conditional on the price of the underlying never exceeding $S = 115$ up to expiry T).
8. Consider a *cliquet option* with pay-off $P = \sum_{i=1}^{n} \left(S_{t_i}/S_{t_{i-1}} - K\right)^+$, where $t_i = i$ denotes the i-th month, and derive a pricing formula in the CRR model with monthly price jumps of the underlying. What is the resulting price of the option in Exercise 7, with $K = 0.05$ and maturity $T = 5$ months? Explain why the CRR model is not particularly well-suited for the pricing of cliquet options.
9. Use the model in Exercise 7 to compute the fair price of a Bermudan put option with maturity $T = 5$ months, possible exercise times $t_i = i$ months ($i = 1, \dots, 5$) and strike price $K = 100$. Hint: since this option can be exercised at every observation time t_i, the value V_t^G of the corresponding European put option, as derived in Section 5.3 by backward substitution, in a particular node G at time t is given by

$$\max\{V_t^G, K - S^G\},$$

where S^G denotes the price of the underlying in this node.

The Black-Scholes Model 6

In the last chapter we introduced a binomial model, which provided an intuitive way for pricing derivatives and finding replicating portfolios. However, the binomial model often oversimplifies the real world, so that in practice one would aim to choose a model setup that better describes reality. In this chapter we will discuss a continuous-time model which is broadly considered today the classical model of mathematical finance.

S_n shall denote the price of a stock at the end of the n-th trading day. The daily *return* from day n to day $n+1$ is then given as the relative price change $(S_{n+1} - S_n)/S_n = S_{n+1}/S_n - 1$. In practice, it is common to work with *log returns* $\log(S_{n+1}/S_n)$ instead.[1] This is mostly due to the fact that the log return of a period of k days can simply be computed by adding up the daily log returns:

$$\log(S_k/S_0) = \log(S_1/S_0) + \cdots + \log(S_k/S_{k-1}).$$

The daily log returns could then again be interpreted as the sum of the hourly log returns, etc.[2]

Under the assumption that log returns over disjoint time intervals are stochastically independent, and that log returns over disjoint equidistant time intervals are identically distributed, the Central Limit Theorem of Probability Theory implies that log returns are close to normally distributed. This is due to their property of being the sum of many small independent and identically distributed ('i.i.d.') random variables of finite variance.[3]

We now search for a stochastic market model that is defined in continuous time and in which log returns over arbitrary time intervals are normally distributed. The

[1] When writing *log* in this book, we refer to the natural logarithm, i.e. with base e.

[2] For small changes in the stock price, the log return and the return will provide very similar results. This can be understood from the Taylor series expansion $\log x = (x-1) - (x-1)^2/2 + \cdots$.

[3] Chapter 8 will test this assumption against market data.

H. Albrecher et al., *Introduction to Quantitative Methods for Financial Markets*, Compact Textbooks in Mathematics, DOI 10.1007/978-3-0348-0519-3_6,

so-called *geometric Brownian motion*, which we will now examine in detail, satisfies these desired properties.

6.1 Brownian Motion and Itô's Lemma

The following process is a key building block of Stochastic Analysis and Financial Mathematics.

Definition (Brownian Motion). A *Brownian motion*[4] (also: *Wiener process*) is a random function $(W_t : t \in \mathbb{R})$ with the following properties:

(i) It holds with probability 1 that $W_0 = 0$ and W_t is a continuous function of t.

(ii) For all $t \geq 0$ and $h > 0$, the increment $W_{t+h} - W_t$ is normally distributed with mean 0 and variance h, i.e.

$$W_{t+h} - W_t \sim N(0, h). \tag{6.1}$$

(iii) For all n and times $t_0 < t_1 < \cdots < t_{n-1} < t_n$, the increments $W_{t_j} - W_{t_{j-1}}$ $(j = 1, \dots, n)$ are stochastically independent.

In particular, this implies that for all t, W_t is normally distributed with mean 0 and variance t. Furthermore, the increments of W_t are *stationary*, i.e. the distribution of $W_{t+h} - W_t$ is independent of t.

In the financial context, Louis Bachelier suggested the use of Brownian motion (cf. Figure 6.1) for the modeling of stock price movements already in 1900.[5]

Property (6.1) yields

$$\Delta W_t := W_{t+\Delta t} - W_t \sim \epsilon \sqrt{\Delta t},$$

with ϵ being a standard normally distributed random variable.

[4]The Brownian motion is named after the Scottish botanist Robert Brown (1773–1858), who first observed the zigzag movements of pollen corns through a light microscope in 1827. The true discovery of these movements is, however, often attributed to the Dutch botanist Jan Ingenhousz (1730–1799), who already described them in 1785 when examining the movements of coal dust on alcohol. The correct physical interpretation of these movements as a consequence of uncoordinated collisions of continuously moving atoms and molecules was first provided by Albert Einstein and Marian Smoluchowski in 1906.

[5]Louis Bachelier (1870–1946) is seen as one of the founders of modern financial mathematics. In his dissertation on the "Théorie de la spéculation" (*Theory of Speculation*) at Sorbonne university in Paris in 1900, he introduced a mathematical framework to deal with Brownian motion five years before Albert Einstein's related work. Bachelier's mentor Henri Poincaré had his work published in the prestigious journal "Annales Scientifiques de l'École Normale Supériore". Despite Andrey Kolmogorov praising this publication, it temporarily disappeared from scientific discussions, and only regained attention when economist Paul Samuelson rediscovered it in Harvard's library and further developed Bachelier's ideas.

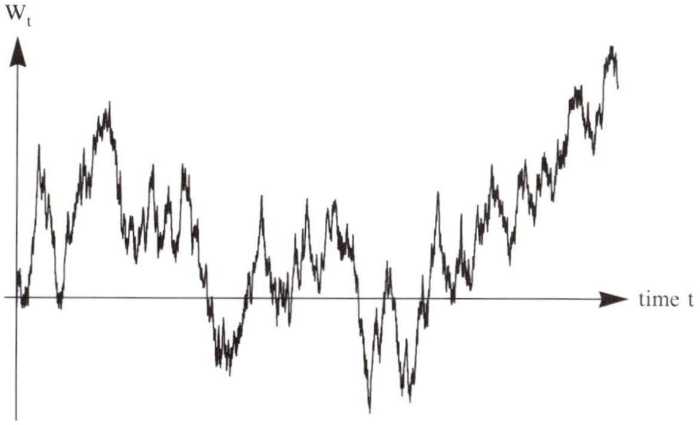

Fig. 6.1 Sample path of a Brownian motion

In this way we can consider infinitesimal increments dW_t (ΔW_t for $\Delta t \to 0$) as

$$dW_t \sim \epsilon\sqrt{dt}. \tag{6.2}$$

The *stochastic integral*[6] for suitable functions f is then defined as

$$\int_0^T f(t)\,dW_t := \lim_{n\to\infty}\sum_{j=1}^{n} f(t_{j-1})\,(W_{t_j} - W_{t_{j-1}}) \quad \text{with} \quad t_j = \frac{jT}{n}. \tag{6.3}$$

As opposed to classical Riemann integration, it is essential in this definition that the function f is evaluated at the left end t_{j-1} of every interval $[t_{j-1}, t_j)$. For $f \equiv 1$ it then immediately follows that $\int_0^T dW_t = W_T$. Based on moment properties of the normal distribution it can be shown that (see Exercise 1)

$$\lim_{n\to\infty}\mathbb{E}\left[\left(\sum_{j=1}^{n}(W_{t_j} - W_{t_{j-1}})^2 - T\right)^2\right] = 0. \tag{6.4}$$

Upon setting $f \equiv 1$ in (6.3), one can interpret the above formula as

$$\int_0^T (dW_t)^2 = T,$$

or further

$$(dW_t)^2 = dt. \tag{6.5}$$

[6] Integrating over a stochastic process is formally challenging, and in the following we will only argue heuristically, which will be sufficient for our purposes. Check the references at the end of this chapter for sources discussing stochastic integrals in a mathematically rigorous way.

From this perspective, it can be argued that the stochastic nature of $(dW_t)^2$ can be 'neglected' in the first order. Strictly speaking, this is only true for the above mean squared deviations, however, for all applications in this book the heuristic version (6.5) will be correct. Based on the infinitesimal increment dW_t, an entire class of stochastic processes can now be defined as

$$dX_t = \mu(X_t, t)\, dt + \sigma(X_t, t)\, dW_t, \tag{6.6}$$

where $\mu(x,t) : \mathbb{R} \times \mathbb{R}^+ \to \mathbb{R}$ and $\sigma(x,t) : \mathbb{R} \times \mathbb{R}^+ \to \mathbb{R}^+$ are sufficiently smooth deterministic functions. Equations of the above type are called *stochastic differential equations* (SDEs), and stochastic processes X_t that satisfy (6.6) will play a major role in the modeling of stock prices (and are called *Itô processes*[7] or *diffusion processes*). The stochastic differential equation (6.6) can be seen as a representation of[8]

$$X_T = X_0 + \int_0^T \mu(X_t, t)\, dt + \int_0^T \sigma(X_t, t)\, dW_t.$$

It will subsequently be useful to have an expression for the dynamics of some sufficiently smooth function $f(X_t, t)$ when the dynamics of X_t are given by (6.6). This is provided by the following key result of stochastic analysis.

Theorem 6.1 (Itô's Lemma). *Let $f(x,t)$ be a sufficiently smooth function and let the stochastic process X_t be defined by* (6.6)*. Then, with probability 1,*

$$df(X_t, t) = \left(\frac{\partial f}{\partial X_t} \mu(X_t, t) + \frac{\partial f}{\partial t} + \frac{1}{2} \frac{\partial^2 f}{\partial X_t^2} \sigma^2(X_t, t) \right) dt + \frac{\partial f}{\partial X_t} \sigma(X_t, t)\, dW_t. \tag{6.7}$$

Proof sketch. The infinitesimal increment $df(X_t, t)$ can be developed by a second-order Taylor expansion as

$$\begin{aligned} df(X_t, t) &= f(X_{t+dt}, t + dt) - f(X_t, t) \\ &= \frac{\partial f}{\partial X_t} dX_t + \frac{\partial f}{\partial t} dt + \frac{1}{2} \frac{\partial^2 f}{\partial X_t^2} (dX_t)^2 \\ &\quad + \frac{\partial^2 f}{\partial X_t \partial t} dX_t\, dt + \frac{1}{2} \frac{\partial^2 f}{\partial t^2} (dt)^2 + \cdots . \end{aligned}$$

[7] Kiyoshi Itô (1915-2008) developed, amongst other things, the concept of stochastic integration in a mathematically rigorous way. As an appreciation of his work in the field of Stochastic Analysis, he received the Gauss prize of the International Mathematical Union in 2006.

[8] Note that also the integrand on the right is a stochastic process.

Substituting here (6.6) and applying (6.5) yields[9]

$$df(X_t,t) = \left(\frac{\partial f}{\partial X_t}\mu(X_t,t) + \frac{\partial f}{\partial t} + \frac{1}{2}\frac{\partial^2 f}{\partial X_t^2}\sigma^2(X_t,t)\right)dt$$
$$+\frac{\partial f}{\partial X_t}\sigma(X_t,t)\,dW_t + o(dt),$$

which ultimately leads to (6.7). □

Due to property (6.5), already the first-order differential $df(X_t)$ contains an additional term that includes the second derivative of f. Itô's formula (6.7), which is the Stochastic Analysis counterpart to the chain rule for deterministic derivatives, also allows to derive product and quotient rules (cf. Exercises 5 and 6).

6.2 The Black-Scholes Model

Let S_t be the price of a stock at time t. Bachelier's model, which describes S_t by a Brownian motion, shows some disadvantages, including the possibility of the stock prices dropping negative. In 1965, P. Samuelson[10] suggested to use a modification, namely the *geometric Brownian motion*

$$dS_t = S_t(\mu\,dt + \sigma\,dW_t) \quad (\mu,\sigma \text{ constant}) \tag{6.8}$$

for the modeling of stock prices. The parameters μ and σ give the *expected rate of return (drift)* and the *volatility* of the stock price process, respectively.[11]

It is obvious that this process S_t is a special case of an Itô process with parameters $\mu(S_t,t) = S_t\mu$ and $\sigma(S_t,t) = S_t\sigma$, for fixed σ and μ. Applying Itô's Lemma (6.7) to $f(S_t,t) = \log S_t$ leads to (see Exercise 2)

$$d(\log S_t) = (\mu - \sigma^2/2)\,dt + \sigma\,dW_t, \tag{6.9}$$

and together with $W_0 = 0$, one obtains for all $T > 0$

$$\log S_T - \log S_0 = (\mu - \sigma^2/2)\,T + \sigma(W_T - W_0) = (\mu - \sigma^2/2)\,T + \sigma W_T.$$

[9]We write $f(x) = o(x)$ if $f(x)/x \to 0$ for $x \to 0$.

[10]Paul Samuelson (1915–2009) received the 1970 Nobel Memorial Prize in Economic Sciences for his results in various fields of economics. He also introduced the terms 'European' and 'American' in the context of derivatives.

[11]In practice, the volatility parameter is often stated as a percentage number, e.g. a volatility of 25% corresponds to $\sigma = 0.25$.

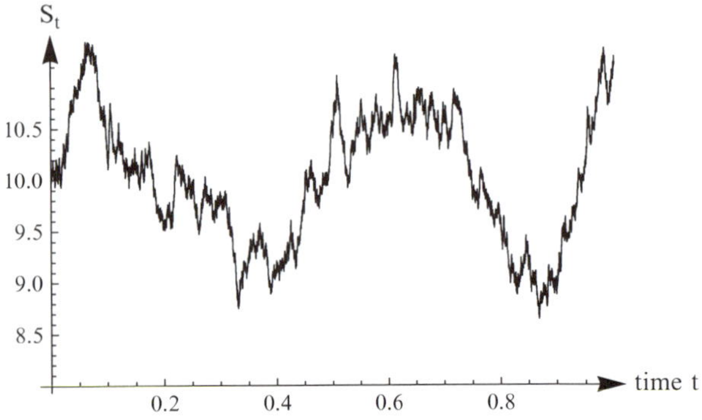

Fig. 6.2 Sample path of a geometric Brownian motion with parameters $\mu = 0.1$, $\sigma = 0.25$ and $S_0 = 10$

As W_T is normally distributed, it follows that

$$\log S_T \sim N\Big(\log S_0 + (\mu - \sigma^2/2)\,T, \sigma^2 T\Big).$$

The geometric Brownian motion therefore models the stock price at time T as a log-normal random variable. Consequently, S_T is positive and we have

$$\mathbb{E}(S_T) = S_0 e^{\mu T} \quad \text{and} \quad \mathrm{Var}(S_T) = S_0^2 e^{2\mu T}(e^{\sigma^2 T} - 1). \tag{6.10}$$

Figure 6.2 depicts a sample path of the process S_t. Discretizing equation (6.8) allows for a graphical interpretation of the parameters μ and σ as

$$\frac{\Delta S_t}{S_t} = \mu \Delta t + \sigma \epsilon \sqrt{\Delta t} \sim N(\mu\,\Delta t, \sigma^2 \Delta t),$$

with $\Delta S_t = S_{t+\Delta t} - S_t$ being the change of the stock price S_t over a small interval Δt and $\epsilon \sim N(0,1)$.

Recall that $\Delta S_t / S_t$ is the return of the stock, which consists of an expected return component $\mu \Delta t$ and a random (normally distributed) component $\sigma \epsilon \sqrt{\Delta t}$. The standard deviation of the return over Δt is hence given by $\sigma \sqrt{\Delta t}$. Note that, in contrast to Bachelier's model, the return dS_t / S_t is now independent of the stock price S_t. Intuitively this is a desirable property for a model.

F. Black, M. Scholes and R. Merton (1973)[12] were first to show how stochastic methods could be used to derive the value of options in the above model, which is why the model is commonly referred to as the *Black-Scholes* or *Black-Merton-Scholes* model.

[12]Myron Scholes (1941–) and Robert Merton (1944–) were awarded the 1997 Nobel Memorial Prize in Economic Sciences for their work on a new method for valuing derivative instruments. Fischer S. Black (1938-1995) had already died by then.

The *Black-Scholes model* is a continuous-time model of financial markets with finite time horizon T, which satisfies the assumptions as lined out in Section 3.1 and involves two financial goods: a risk-free account paying interest at some fixed (continuously compounded) interest rate r (note that negative account balances are permitted and describe borrowings) and a risky asset (such as a stock). The price development of the risky asset in this model is then described by a geometric Brownian motion (6.8) with constant drift μ and constant volatility σ.

6.3 Key Takeaways, References and Exercises

Key Takeaways

After working through this chapter you should understand and be able to explain the following terms and concepts:

- The advantage of modeling log returns $\log S_{n+1}/S_n$
- Brownian motion: definition, $(dW_t)^2 = dt$, Itô process
- Itô's Lemma for $df(X_t, t)$
- Black-Scholes model: geometric Brownian motion, distribution of $\log S_T$, $\mathbb{E}[S_T]$, $\mathrm{Var}[S_T]$, distribution of $\Delta S_t/S_t$ in its discretized version (e.g. for simulation)

References

For a mathematically rigorous introduction to Brownian motion and Stochastic Analysis, the interested reader is referred to the books Karatzas & Shreve [45], Rogers & Williams [65], or Øksendal [60]. A nicely presented discussion of the historic development of Itô Calculus is given in Schachermayer & Teichmann [67]. The original articles Black & Scholes [9] and Merton [57] are also fairly accessible to readers who prefer to avoid mathematical technicalities.

Exercises

1. Show that

$$\lim_{n\to\infty} \mathbb{E}\left[\left(\sum_{j=1}^{n}(W_{t_j} - W_{t_{j-1}})^2 - T\right)^2\right] = 0$$

   as in (6.4), which was the crucial relation to intuitively show $(dW_t)^2 = dt$. (Hint: Expand the square inside the expectation, and cleverly use the facts that $\mathbb{E}[(W_{t_j} - W_{t_{j-1}})^2] = t_j - t_{j-1}$ and that $\mathbb{E}[X^4] = 3\sigma^4$ for $X \sim N(0, \sigma^2)$.)
2. Prove that $d(\log S_t) = (\mu - \sigma^2/2)\,dt + \sigma\,dW_t$ for a geometric Brownian motion S_t (as in (6.9)).
3. Use the moment generating function of the normal distribution to show that the mean and the variance of S_T in the geometric Brownian motion model are indeed given by (6.10).
4. Use Itô's formula to prove that $d(W_t^2) = dt + 2W_t\,dW_t$ holds and, hence, that

$$\int_0^T W_t \, dW_t = \frac{1}{2} W_T^2 - \frac{T}{2}. \tag{6.11}$$

5. Prove the product rule

$$d\left(f(X_t,t)\cdot g(X_t,t)\right) = f(X_t,t)\cdot dg(X_t,t) + g(X_t,t)\cdot df(X_t,t) + df(X_t,t)\cdot dg(X_t,t)$$

by applying Itô's formula. (Hint: use $(dW_t)^2 = dt$.)

6. Prove the quotient rule

$$d\left(f(X_t,t)/g(X_t,t)\right) = \frac{g(X_t,t)\cdot df(X_t,t) - f(X_t,t)\cdot dg(X_t,t) - df(X_t,t)\cdot dg(X_t,t)}{(g(X_t,t))^2}$$

by using Itô's formula. (Hint: use $(dW_t)^2 = dt$.)

7. Let the price S_t of a stock be given by a geometric Brownian motion with volatility $\sigma = 0.35$ and an annual expected return of 16 % ($\mu = 0.16$). Today's stock price shall be $S_0 = 100$ EUR. Compute the probability that a
   (a) European call option
   (b) European put option
   on this stock, with strike price $K = 90$ EUR and maturity $T = 6$ months, will be exercised?

The Black-Scholes Formula 7

For the Black-Scholes model, as introduced in the last chapter, we can now derive the no-arbitrage price of a European-style option – the so-called *Black-Scholes formula*. In Section 7.1, we will discuss a direct approach to obtaining the Black-Scholes formula as the solution of a partial differential equation. In Section 7.2, we will see that the Black-Scholes model can also be interpreted as a limit of the discrete Cox-Ross-Rubinstein model (cf. Section 5.3), and that the Black-Scholes price of a European call can hence be derived as a limit of the corresponding price in the discrete setup.

7.1 The Black-Scholes formula from a PDE

Consider the Black-Scholes model, where $C(S,t)$ is the price at time t of a European call option with strike K and maturity T, and the underlying stock price is $S = S_t$. Suppose a bank sells this option at time t and charges a premium consisting of the fair premium $C(S,t)$ plus some fee as profit from the trade. The value of the liability under the option contract will now be reported on the bank's books. As the price of the underlying moves (stochastically), the fair value of the liability does so too. Typically, the bank prefers to eliminate balance sheet fluctuations and to lock in the profit from the trade, so that it will now add *hedge positions* to its portfolio to cancel the value fluctuations from the option. In the Black-Scholes model, assume a portfolio that consists at time t of (i) one call option pricing at $C(t,T)$, (ii) $-\Delta_t$ units in the underlying stock (Δ_t must still be determined) and (iii) $C(t,T) - \Delta_t S_t$ monetary units in the risk-free account (negative values refer to a borrowing). The value of this portfolio at time t shall be π_t, and the portfolio construction implies $\pi_0 = 0$.

Consider now the value change of the single portfolio positions over the infinitesimal time interval $[t, t+dt)$:

- the underlying: its value change is captured by dS. Hence, the value change from holding Δ_t units of the underlying is given by $\Delta_t \cdot dS$

H. Albrecher et al., *Introduction to Quantitative Methods for Financial Markets*, Compact Textbooks in Mathematics, DOI 10.1007/978-3-0348-0519-3_7,

- the risk-free account: interest is paid/received so that the change to this position is given by $(C(S,t) - \Delta_t \cdot S)rdt$
- the call option (short): the change is described by $dC(S,t) = C(S + dS, t + dt) - C(S,t)$. Itô's Lemma (6.7) yields

$$dC(S,t) = \left(\frac{\partial C}{\partial S} \mu S + \frac{\partial C}{\partial t} + \frac{1}{2} \frac{\partial^2 C}{\partial S^2} \sigma^2 S^2 \right) dt + \frac{\partial C}{\partial S} \sigma S \, dW_t.$$

Aggregating over the single positions, the change of the portfolio value over $[t, t + dt)$ is then given by

$$\begin{aligned} d\pi_t &= \Delta_t \cdot dS + (C(S,t) - \Delta_t \cdot S) r \, dt - dC(S,t) \\ &= \left(\Delta_t - \frac{\partial C}{\partial S} \right) dS + (C(S,t) - \Delta_t \cdot S) \, r \, dt - \left(\frac{\partial C}{\partial t} + \frac{1}{2} \frac{\partial^2 C}{\partial S^2} \sigma^2 S^2 \right) dt \end{aligned}$$

Note that the stochastic component dS in the above sum can be eliminated if

$$\Delta_t = \frac{\partial C}{\partial S},$$

so that the no-arbitrage principle requires the deterministic component to vanish as well (due to $\pi_0 = 0$). This yields the condition

$$\left(\frac{\partial C}{\partial t} + \frac{1}{2} \frac{\partial^2 C}{\partial S^2} \sigma^2 S^2 \right) dt = r \left(C(S,t) - \frac{\partial C}{\partial S} \cdot S \right) dt,$$

whence

$$\frac{\partial C}{\partial t} + \frac{\sigma^2 S^2}{2} \frac{\partial^2 C}{\partial S^2} + r S \frac{\partial C}{\partial S} - r C = 0. \tag{7.1}$$

This linear parabolic *partial differential equation* (PDE) for $C(S,t)$ is often referred to as *Black-Scholes differential equation*. Note that the pay-off structure of the call option has not yet entered the calculations, so that the Black-Scholes differential equation will be satisfied for arbitrary European-style derivatives. The pay-off of the to-be-priced derivative is reflected in the *boundary condition* at $t = T$. In the case of a European call with maturity T and strike price K, the boundary condition at $t = T$ is simply given by

$$C(S,T) = (S - K)^+. \tag{7.2}$$

Moreover, the no-arbitrage bounds for the call price in Section 4.1 imply that

$$C(0,t) = 0, \quad \lim_{S \to \infty} C(S,t)/S = 1 \tag{7.3}$$

must hold for all $t \in [0, T]$.

Equation (7.1), under the conditions (7.2) and (7.3), can then be solved analytically and uniquely (see Exercise 1) to give the well-known Black-Scholes formula.

Theorem 7.1 (Black-Scholes formula). *In the Black-Scholes model, the price of a European call option defined by the parameters K, T, r, σ and S_0 is given by*

$$C_0 = S_0\Phi(d_+) - e^{-rT} K\Phi(d_-) \tag{7.4}$$

with

$$d_\pm = \frac{\log(S_0/K) + (r \pm \frac{1}{2}\sigma^2)T}{\sigma\sqrt{T}},$$

where $\Phi(x)$ is the cumulative distribution function of the standard normal distribution.

7.2 The Black-Scholes Formula as Limit in the CRR-Model

The Black-Scholes formula for the price of a European call option can also be derived through interpreting the CRR model as a discrete approximation of the price process of the underlying. One can then take this approximation to the limit, and the CRR pricing formula will converge to the Black-Scholes formula.

We start by modeling the price of the underlying by a CRR model, with initial price S_0, and we allow the price on $[0, T]$ to change at n discrete times $\{0, h, 2h, \ldots, nh\} \subset [0, T]$, with $h = T/n$. The idea is now to let $n \to \infty$. To achieve this, the jump parameters $a(n)$ and $b(n)$ in the CRR model have to be chosen as suitable functions of n. Define

$$1 + b(n) = e^{rT/n}\, e^{+\sigma\sqrt{T/n}} \quad \text{and} \quad 1 + a(n) = e^{rT/n}\, e^{-\sigma\sqrt{T/n}}, \tag{7.5}$$

with $\sigma > 0$ fixed such that we attain the Black-Scholes model in the limit. Substituting (7.5) into (5.8) leads to the risk-neutral probability

$$q(n) = \frac{1 - e^{-\sigma\sqrt{T/n}}}{e^{+\sigma\sqrt{T/n}} - e^{-\sigma\sqrt{T/n}}}.$$

The returns R_i of the price of the underlying over the time interval $(iT/n, (i+1)T/n)$, with $i = 1, \ldots, n$, under the risk-neutral probability measure $\mathbb{Q}$ are therefore given by

$$R_i(n) = \begin{cases} e^{rT/n}\, e^{+\sigma\sqrt{T/n}} - 1 & \text{with probability } q(n), \\ e^{rT/n}\, e^{-\sigma\sqrt{T/n}} - 1 & \text{with probability } 1 - q(n). \end{cases}$$

Consider

$$Z(n) := \sum_{i=1}^{n} \log\left(\frac{R_i(n)+1}{e^{rT/n}}\right)$$

and note that

$$S_T = S_0 \cdot e^{-rT} \prod_{i=1}^{n}(R_i(n)+1) = S_0 \cdot \exp\left(\sum_{i=1}^{n} \log\left(\frac{R_i(n)+1}{e^{rT/n}}\right)\right) = S_0 \cdot e^{Z(n)}.$$

Recalling the results of Chapter 5, the call price in this model can be written as

$$C_0(n) = \mathbb{E}^{\mathbb{Q}}\left[\left(S_0\, e^{Z(n)} - e^{-rT} K\right)^{+}\right]. \tag{7.6}$$

It remains to show that $C_0(n)$ as above converges to the price in the Black-Scholes model when $n \to \infty$.

Lemma (Central Limit Theorem (CLT)). *Let* $(Y_k(n))_{k \le n}$ *be a sequence of independent and identically distributed (i.i.d.) random variables with means* $\mu(n)$*, such that* $(n\mu(n)) \overset{n\to\infty}{\longrightarrow} \mu < \infty$*, and variances* $\sigma^2/n + o(1/n)$*. It follows that*[1]

$$Z(n) = \sum_{k=1}^{n} Y_k(n) \overset{d}{\longrightarrow} Z,$$

with $Z \sim N(\mu, \sigma^2)$*, i.e.,* Z *is normally distributed with mean* μ *and variance* σ^2*.*

To apply this version of the Central Limit Theorem, one has to verify that the mean and the variance of

$$Y_i(n) := \log\left(\frac{R_i(n)+1}{e^{rT/n}}\right) = \begin{cases} \sigma\sqrt{T/n} & \text{with probability } q(n) \\ -\sigma\sqrt{T/n} & \text{with probability } 1-q(n) \end{cases}$$

satisfy the conditions of the lemma. For the mean, it holds that

$$\mathbb{E}[Y_i(n)] = \sigma\sqrt{T/n}\, q(n) - \sigma\sqrt{T/n}\,(1-q(n)) = (2q(n)-1)\,\sigma\sqrt{T/n}.$$

[1] The notation $\overset{d}{\longrightarrow}$ indicates that the distribution of $Z(n)$ converges to the distribution of Z. Under the assumptions of this version of the Central Limit Theorem, the so-called *Lindeberg condition*, $\lim_{n\to\infty} \frac{1}{n\sigma^2(n)} \sum_{k=1}^{n} \mathbb{E}\left[Y_k(n)^2 \mathbf{1}_{\{|Y_k(n)|>\epsilon\}}\right]$ for all $\epsilon > 0$, is fulfilled, under which the result holds. For details consult Feller [31].

Hence, it remains to show that $2q(n)-1$ is of order $1/\sqrt{n}$. This can be done by a Taylor series expansion of $2q(n)-1$ with respect to $\sqrt{T/n}$ (cf. Exercise 3):[2]

$$\begin{aligned}2q(n)-1 &= 1-2(1-q(n)) = 1-2\,\frac{e^{+\sigma\sqrt{T/n}}-1}{e^{+\sigma\sqrt{T/n}}-e^{-\sigma\sqrt{T/n}}}\\ &= -\frac{\sigma\sqrt{T}}{2}\,\frac{1}{\sqrt{n}} + O(1/n).\end{aligned}$$

For the variance, on the other hand, one can write

$$\begin{aligned}\mathrm{Var}[Y_i(n)] &= \mathbb{E}[Y_i(n)^2]-(\mathbb{E}[Y_i(n)])^2 = \frac{\sigma^2 T}{n} - \Big(-\frac{\sigma^2 T}{2n} + o(1/n)\Big)^2\\ &= \frac{\sigma^2 T}{n} + o(1/n).\end{aligned}$$

Therefore, the requirements of the CLT are satisfied and it follows that $Z(n) \xrightarrow{d} Z$ with $Z \sim N(-\frac{1}{2}\sigma^2 T, \sigma^2 T)$. The CRR model therefore converges to the Black-Scholes model and the price of the call $C_0(n)$ converges to[3]

$$C_0 = \mathbb{E}\Big[\Big(S_0\, e^{Z} - e^{-rT} K\Big)^+\Big].$$

Upon standardization of Z it holds that $X = (1/\sigma\sqrt{T})(Z+\frac{1}{2}\sigma^2 T) \sim N(0,1)$, or conversely, that $Z = -\frac{1}{2}\sigma^2 T + \sigma\sqrt{T}X$ for $X \sim N(0,1)$. The limit of $C_0(n)$ can then be derived as

$$\begin{aligned}C_0 &= \int_{-\infty}^{\infty}\Big(S_0 e^{-\frac{1}{2}\sigma^2 T+\sigma\sqrt{T}x} - e^{-rT}K\Big)^+ \frac{e^{-\frac{1}{2}x^2}}{\sqrt{2\pi}}\,dx\\ &= S_0\int_{\gamma}^{\infty} e^{-\frac{1}{2}\sigma^2 T} e^{\sigma\sqrt{T}x-\frac{1}{2}x^2}\frac{dx}{\sqrt{2\pi}} - Ke^{-rT}(1-\Phi(\gamma))\\ &= S_0\int_{\gamma}^{\infty} e^{-\frac{(x-\sigma\sqrt{T})^2}{2}}\frac{dx}{\sqrt{2\pi}} - Ke^{-rT}(1-\Phi(\gamma))\\ &= S_0\Big(1-\Phi(\gamma-\sigma\sqrt{T})\Big) - Ke^{-rT}(1-\Phi(\gamma)),\end{aligned}$$

---

[2] $f(n) = O(g(n))$ means that there exist $M, n_0 > 0$, such that $f(n) \le M \cdot g(n)$ for all $n \ge n_0$. The notation $f(n) = o(g(n))$, on the other hand, is used if $f(n)/g(n) \to 0$ for $n \to \infty$.

[3] Formally, pulling the limit inside the expectation (i.e. the integral) as in (7.6) requires further justification. It can either be proven that $(S_0\, e^{Z(n)} - e^{-rT}K)^+$ is *uniformly integrable*, or one can first derive the formula for a put option (in which case the interchange is justified by *dominated convergence*) and subsequently apply the put-call parity.

where $\Phi(x)$ is the cumulative distribution function of the standard normal distribution and

$$\gamma = \frac{\log(K/S_0) + (\frac{1}{2}\sigma^2 - r)T}{\sigma\sqrt{T}}.$$

Rewriting this formula yields the Black-Scholes formula (7.4).

7.3 Discussion of the Formula, Hedging

Replacing T by $T-t$ and S_0 by S_t in (7.4) gives the value C_t of the option at time t. In this case one can see the option as a contract that is entered at time t and has a maturity of $T-t$, i.e.

$$C_t = S_t\Phi(d_{t+}) - e^{-r(T-t)}K\Phi(d_{t-}), \tag{7.7}$$

with

$$d_{t\pm} = \frac{\log(S_t/K) + (r \pm \frac{1}{2}\sigma^2)(T-t)}{\sigma\sqrt{T-t}}.$$

Applying the put-call parity to (7.7) leads to the Black-Scholes price P_t of a European put option with the same parameters,

$$P_t = Ke^{-r(T-t)}\Phi(-d_{t-}) - S_t\Phi(-d_{t+}). \tag{7.8}$$

Let us now further examine the behavior of the price C_t (a similar line of argument will hold true for P_t):

- for $S_t \to \infty$ it follows that $d_{t\pm} \to \infty$ in (7.7), so that $\Phi(d_{t\pm}) \to 1$ and C_t converges to $S_t - Ke^{-r(T-t)}$. The option can then be seen as a forward contract with strike price K, since the long position will execute its right to buy the underlying at time T 'with certainty'.
- for $\sigma \to 0$ we find $d_{t\pm} \to \infty$, so that the underlying will in this case behave like a risk-free bond or cash account.
- For $T-t \to 0$ (i.e. as the maturity of the option is approached) and $S_t > K$, one observes that $d_{t\pm} \to \infty$ and $e^{-r(T-t)} \to 1$. Hence, C_t tends to $S_t - K$. For the case where $S_t < K$, we find $\log(S_t/K) < 0$, so that $d_{t\pm} \to -\infty$ and $C_t \to 0$. As expected, this implies that $C_t \to (S_T - K)^+$ for $t \to T$.

One can read the following hedging strategy of formula (7.7).

Theorem 7.2 (Replicating portfolio in the Black-Scholes model). *At time t, the value C_t of the option can be reproduced by a portfolio of*

$$\theta_t^0 = -Ke^{-r(T-t)}\Phi(d_{t-}), \qquad \theta_t^1 = \Phi(d_{t+})$$

units in the risk-free account and the underlying stock, respectively.

Remark 7.3. The option value (7.4) depends on the risk-free interest rate r and the volatility σ of the price of the underlying, but *not* on the drift μ of the price of the underlying. The derivation of the Black-Scholes formula only required the drift μ to be constant, but the concrete value of μ does not affect the value of the option. To put it differently, two market participants can agree on the price of the option although they may disagree on the expected return of the underlying.

Remark 7.4. As in the CRR model, one can determine a (unique) risk-neutral measure $\mathbb{Q}$ which can be applied to the pricing of derivatives by computing the discounted expected pay-off of the derivative under $\mathbb{Q}$. It can be shown that the distribution of the price S of the underlying under the measure $\mathbb{Q}$ is given by

$$dS_t = S_t(rdt + \sigma dW_t^{\mathbb{Q}}),$$

with $W^{\mathbb{Q}}$ being a Brownian Motion (under $\mathbb{Q}$). This means that the probabilities of particular pay-offs in the Black-Scholes model are re-weighted by changing the mean return from μ to r.[4]

If dividend payments should also be considered in the Black-Scholes setup, one could define some dividend rate q, such that dividends are paid at the constant rate q and proportional to the price of the stock S_t. The constant dividend rate assumption is not particularly realistic for single stocks,[5] however, it has the advantage that the model under $\mathbb{Q}$ only has to be slightly adapted. In particular, the modification is of the form

$$dS_t = S_t((r-q)dt + \sigma dW_t^{\mathbb{Q}}).$$

Note that the Black-Scholes model can also be applied to the modeling of currency exchange rates under the assumption that one can identify risk-free interest rates r_d and r_f in the *domestic* and the *foreign* currency. The exchange rate S_t under the risk-neutral measure $\mathbb{Q}$ is then given by (see Exercise 5)

$$dS_t = S_t((r_d - r_f)dt + \sigma dW_t^{\mathbb{Q}}).$$

---

[4]Heuristically, this can be seen by taking the limit in the CRR model. For a formal proof using *Girsanov's Theorem* consult the references at the end of the chapter.

[5]This assumption is better suited for indices.

7.4 Delta-Hedging and the 'Greeks'

When deriving the Black-Scholes differential equation in Section 7.1, we saw that the risk of a call option can be hedged away with a short position of

$$\Delta = \frac{\partial C(S_t, t)}{\partial S_t} \tag{7.9}$$

('*Delta*') units of the underlying (the so-called Δ-*hedge* of the option), i.e. a portfolio of (i) a call and (ii) a short position in Δ units of the underlying behaves like a deterministic investment over $[t, t + dt)$.[6] (7.9) is the *Delta* of the call option, and it changes over time as the prices of the option and the underlying change. This implies that the Δ-hedged portfolio must be adjusted (i.e. *re-hedged*) continuously. Continuous re-hedging is not possible in practice, and re-balancing the portfolio at discrete times will generate a certain *hedging error* (see Section 7.5). The Δ of an option is an important sensitivity measure of the call price with respect to a change in the price S_t. Figure 7.1 plots the Δ of a European call option as function of strike and maturity.

The *Gamma* of an option is defined as

$$\Gamma := \frac{\partial^2 C(S_t, t)}{\partial S_t^2}$$

and measures the sensitivity of Δ to changes in S_t. Hence, Γ is a measure of how often a Δ-hedge has to be adjusted to control the hedging error. For example, if Γ is close to 0, a small change in S_t changes Δ only by little, such that regularly re-balancing the portfolio will not be as important as for larger values of Γ. This implies that one would naturally aim to keep the Γ of a portfolio low.[7]

The sensitivity of the call price to the option maturity is described by *Theta*,

$$\Theta := \frac{\partial C(S_t, t)}{\partial t},$$

and to the volatility parameter by *Vega* (sometimes denoted by ν):

$$\text{Vega} := \frac{\partial C(S_t, t)}{\partial \sigma}.$$

Finally, the sensitivity with respect to the interest rate is measured by *Rho*

$$\rho := \frac{\partial C(S_t, t)}{\partial r}.$$

---

[6] This portfolio is commonly referred to as Δ-*neutral portfolio*.

[7] In practice, lowering Γ could, e.g., be achieved by adding options of different strikes.

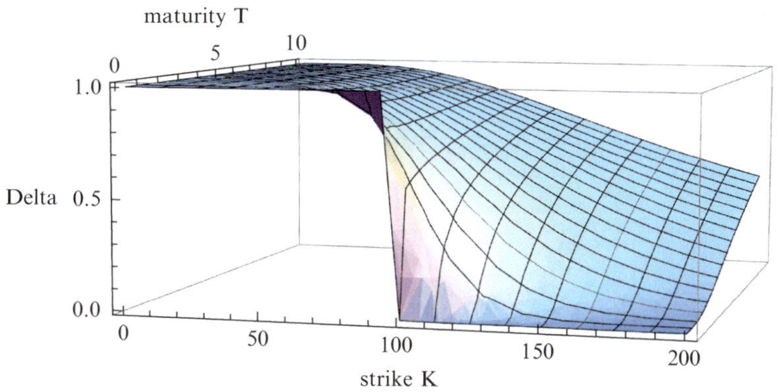

Fig. 7.1 Delta of a European Call in the Black-Scholes model with parameters $r = 0.04$, $\sigma = 0.2$ and $S_0 = 100$ as function of the strike K and the maturity T (in years)

Greek/derivative	European Call	European Put
Δ	$\Phi(d_{t+})$	$-\Phi(-d_{t+})$
Γ	$\phi(d_{t+})(S_t\sigma\sqrt{T-t})^{-1}$	
Θ	$S_t\phi(d_{t+})\sigma(2\sqrt{T-t})^{-1}$ $-rKe^{-r(T-t)}\Phi(d_{t-})$	$S_t\phi(d_{t+})\sigma(2\sqrt{T-t})^{-1}$ $+rKe^{-r(T-t)}\Phi(-d_{t-})$
Vega	$S_t\phi(d_{t+})\sqrt{T-t}$	

Fig. 7.2 Greeks for a European call/put option in the Black-Scholes model

These and other hedging parameters are often referred to as 'The Greeks' and they can be useful tools when hedging a portfolio of financial instruments. Figure 7.2 closes this section by listing the four introduced Greeks for put and call options in the Black-Scholes model.[8]

7.5 Does Hedging Work?

Recall some of the assumptions made when determining the Δ-neutral portfolio consisting of a call option and $-\Delta$ units of the underlying:

1. No transaction costs, liquid markets: for large-scale market participants, transaction costs typically play a less significant role (except when re-hedging is applied almost continuously). However, a lack of market liquidity can pose a major issue. During financial downturns, unhedged open positions in option contracts can lead to great volatility in the prices of the underlying as market participants try to exit their positions. This can aggravate losses of other option holders due to the leveraged structure of option contracts (see Section 2.5).

[8] $\phi(x) = \frac{1}{\sqrt{2\pi}}e^{-x^2/2}$ denotes the probability density function of the standard normal distribution.

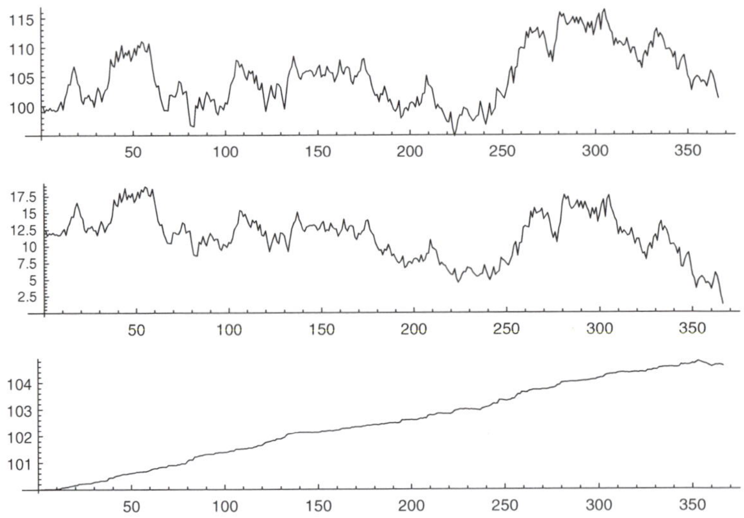

Fig. 7.3 Evolution of (a) the share price (with $\mu = r$), (b) the option value, and (c) the Δ-hedged portfolio value

2. Continuous re-hedging is possible: this will not be the case in practice. Single orders would be too small and transaction costs would be no longer negligible. Possible strategies of re-hedging in practice could be:
   - The portfolio is re-hedged at discrete points in time, at which the portfolio returns to a Δ-neutral state.
   - Re-hedging is performed once the theoretical Δ of the option (which depends on S_t) crosses some predefined bounds.
3. The parameters in the Black-Scholes model are known, so that the fair value of the option and Δ can be computed: even when operating in a Black-Scholes world, the estimates for the volatility may differ.
   - When entering an option contract, a particular σ_1 is implicitly used as Black-Scholes volatility when the buyer and the seller agree on the option premium. σ_1 can simply be based on the estimation of the true parameter, or in an illiquid market it can be inflated as the buyer would be willing to pay an illiquidity premium for entering into the contract (which can be interpreted as increased volatility σ_1).
   - A Δ-hedger might use some other value σ_2, when he aims to implement a self-financing trading strategy.
   - The true parameter σ_3 is typically unknown to market participants.

   For $\sigma_1 = \sigma_2 = \sigma_3$, the performance of a Δ-hedge strategy could look like the one depicted in Figure 7.3 (maturity 365 days, $r = 5\,\%$, $\sigma = 25\,\%$, daily re-hedging of an at-the-money European call option (cf. footnote on p. 92) with $S_0 = K = 100$ and no transaction costs). The (relatively low) deviation of the evolution of the portfolio value from the risk-free investment is due to imperfect

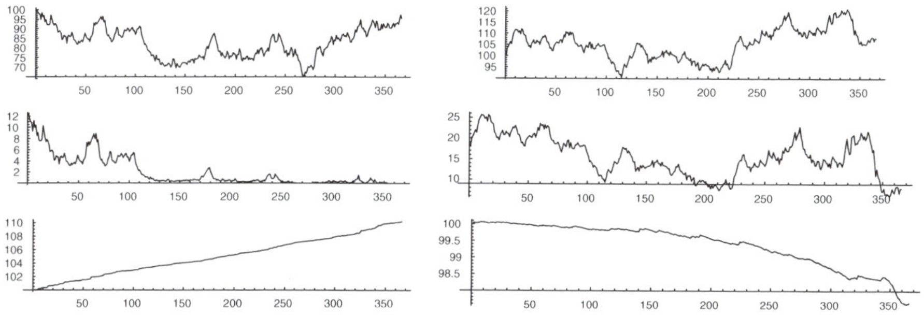

Fig. 7.4 Hedging with under- and overestimated volatility. Left: $\sigma_1 = \sigma_2 = 25\,\%, \sigma_3 = 40\,\%$, Right: $\sigma_1 = \sigma_2 = 40\,\%, \sigma_3 = 25\%$

(daily) hedging. If $\sigma_3 > \sigma_1$, this would resemble a favorable situation for the option buyer, who now profits from positive Vega. Conversely, the option buyer pays too much for the option if $\sigma_3 < \sigma_1$. Figure 7.4 illustrates this situation of over- or underestimating the true volatility.

7.6 Key Takeaways, References and Exercises

Key Takeaways

After working through this chapter you should understand and be able to explain the following terms and concepts:

- Hedge portfolios
- Black-Scholes differential equation, boundary condition defines option pay-off
- Black-Scholes formula as limit of the CRR model, Central Limit Theorem
- Hedging in the Black-Scholes model
- Effect of $S_t \to \infty, \sigma \to 0$ and $T - t \to 0$ on the call option price $C(t, T)$
- Hedging error, the Greeks: Δ, Γ, Θ and Vega
- Hedging issues: transaction costs/liquid markets, continuous re-hedging, knowledge of parameters

References

Various ways of arriving at the Black-Scholes formula are discussed by Elliott & Kopp [26], Baxter & Rennie [4], Duffie [23] or Wilmott [75] (who develops in detail a constructive solution to the Black-Scholes differential equation). Andreasen, Jensen & Paulsen [2] describe as many as eight different ways of obtaining the Black-Scholes formula. For the modeling of exchange rates through a modified Black-Scholes model, consult Garman & Kohlhagen [37].

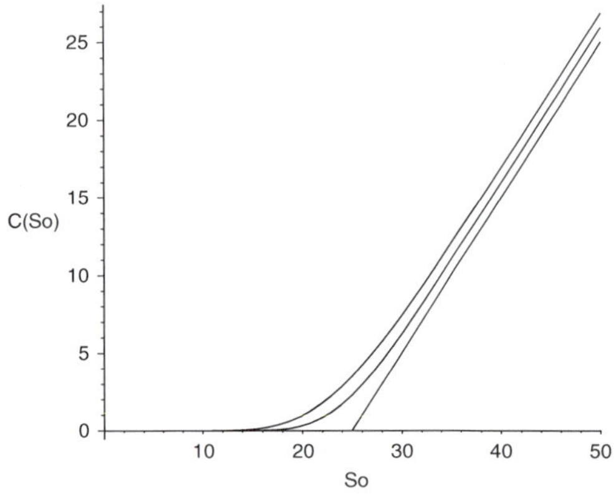

Fig. 7.5 Value of a call option as a function of S_0 for $T = 0.1, 2$

Exercises

1. Check that the Black-Scholes formula (7.4) for a European call option is indeed a solution to equation (7.1) with boundary condition (7.2).
2. Prove the lemma on page 66 by using characteristic functions.
3. Show that the Taylor series expansion of $2q - 1$ with respect to $\sqrt{T/n}$ at 0 is given by $-\sigma/2 \cdot \sqrt{T/n} + o(1/\sqrt{n})$.
4. The price of a stock shall follow a geometric Brownian motion with volatility parameter $\sigma = 0.25$. Assume that $S_0 = 100$ and the risk-free interest rate is $r = 0.04$. Compute the price of a call option and all Greeks, for a maturity of $T = 1$ and a strike price of $K = 105$. Repeat your calculations for a put option, and verify that your obtained call and put price satisfy the put-call parity.
5. Use the hedging ideas of Section 7.1 and the distribution under the risk-neutral risk measure $\mathbb{Q}$ to show that the price C_0 of a European call option with maturity T, strike price K, and a dividend rate of q is given by

$$C_0 = e^{-qT} S_0 \Phi(d_+) - e^{-rT} K \Phi(d_-),$$

with

$$d_\pm = \frac{\log(S_0/K) + (r - q \pm \frac{1}{2}\sigma^2)T}{\sigma\sqrt{T}}.$$

(Hint: when formulating the hedging argument, note that the dividend payments are described by $q\Delta S_t dt$.)

Exercises with Mathematica or UnRisk

8. Consider a Black-Scholes model with $r = 0.04$, volatility $\sigma = 0.18$ and S_0 =20 EUR. Use Mathematica to implement a function that produces the price of a call option with strike price 25 EUR depending on the current price of the underlying. Plot the function for maturities $T_1 = 0.1$ years and $T_2 = 2$ years (cf. Figure 7.5).
9. Verify by symbolically differentiating in Mathematica that the Greeks for plain vanilla European calls and puts in the Black-Scholes model are indeed the ones given in Figure 7.2. Note that the Δ for a European call option corresponds to θ_t^1 on page 69 (although θ_t^1 depends on S_t).
10. Use Mathematica to plot the Greeks as functions of (a) the strike price and (b) the maturity of the call option.

11. Use Mathematica to plot the price of a call option for a range of maturities. Implement a scroll bar that allows to adapt the maturity dynamically.
12. Apply the UnRisk commands `MakeEquity`, `MakyVanillaEquity Option` and `Valuate`, to compare the values of a European and an American put option that is otherwise identical. In what cases do these values only differ by little? When does the difference become more significant?

Stock-Price Models

8

8.1 Shortcomings of the Black-Scholes Model: Skewness, Kurtosis and Volatility Smiles

In Chapter 7 it has been shown that the Black-Scholes model allows to derive explicit formulas for the prices of European call and put options. Having explicit pricing formulas is a great advantage; however, the Black-Scholes model has also been found to not fully explain market prices due to some of its assumptions and properties.

Recall that the Central Limit Theorem provided an intuitive explanation for the assumption of log-normally distributed returns. Empirical studies, however, suggest that stock log returns are not normally distributed in reality. In particular, empirical distributions are typically found to be asymmetric (or: *skewed*) and to have fat tails (i.e. a higher probability of producing extremely low or high outcomes than the normal distribution). Figure 8.1 illustrates this by looking at daily log-returns of the S&P 500 index from January 1999 to December 2008.[1] The empirical distribution is compared to a fitted normal distribution, and we observe that the shapes of the two distributions differ. Calculating the *skewness* and *kurtosis*[2] of the empirical distribution gives −0.156658 and 10.6682, respectively. Both values are significantly different from the respective values 0 and 3 of a normal distribution. The assumption of normally distributed stock log returns hence does not fully reflect reality.

A second shortcoming of the Black-Scholes model is that it cannot explain the so-called *volatility smile* implied by option prices. Examining the Black-Scholes formula (7.4) reveals that all parameters, except for σ, are given by the market

[1] Data source: *Yahoo Finance*.

[2] The skewness coefficient $\mathbb{E}\left[\frac{(X-\mu)^3}{\sigma^3}\right]$ and the kurtosis $\mathbb{E}\left[\frac{(X-\mu)^4}{\sigma^4}\right]$ of a random variable X with mean μ and standard deviation σ are defined as the centralized and scaled 3rd and 4th moments, respectively.

H. Albrecher et al., *Introduction to Quantitative Methods for Financial Markets*, Compact Textbooks in Mathematics, DOI 10.1007/978-3-0348-0519-3_8,

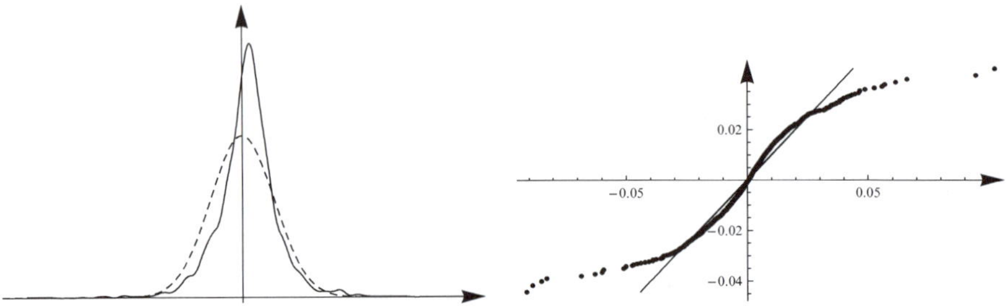

Fig. 8.1 Daily log returns of the S&P 500 index (1999-2008). Left: Empirical probability density function (solid line) and the probability density function of the fitted normal distribution (dashed line). Right: QQ-plot of the empirical vs. the fitted normal distribution

(r and S_0) or the contract specifications (K and T). On the other hand, options exchanges (such as Eurex in Frankfurt or the Chicago Board Options Exchange (CBOE)) provide a relatively liquid market for trading European options, where option prices are determined by demand and supply. A particular option price can then be related to a specific volatility parameter, and one determines the *implied volatility* σ_{imp} of an option price by (uniquely) solving the Black-Scholes formula for σ_{imp} given a market price C^{market} (see Exercise 1),

$$S_t\Phi\left(d_+(\sigma_{\text{imp}})\right) - e^{-r(T-t)}K\Phi\left(d_-(\sigma_{\text{imp}})\right) = C^{\text{market}}(S_t, K, t, T), \tag{8.1}$$

with

$$d_\pm(\sigma_{\text{imp}}) = \frac{\log(S_t/K) + (r \pm \frac{1}{2}\sigma_{\text{imp}}^2)(T-t)}{\sigma_{\text{imp}}\sqrt{T-t}}.$$

If the Black-Scholes model provided an accurate description of reality, σ_{imp} would be the same for all traded European options on the same underlying. However, one generally observes that σ_{imp} is not constant but a function of the strike and the maturity of the option (see Figure 8.2).[3]

It can be concluded that the Black-Scholes model is not able to incorporate all information provided by the markets. Using a Black-Scholes framework therefore ignores some market features, which can in turn lead to errors in the no-arbitrage pricing of new derivatives. Despite its shortcomings, the Black-Scholes model is still popular as a first benchmark in practice due to its simple structure and transparency. In particular, a mathematical model will always simplify reality in an attempt

[3]For example, plotting implied volatility against the strike price for options of the same maturity results in a 'U' shaped, 'smiling' curve, which is why this phenomenon is called *volatility smile*. The 'smile' is often fairly distorted in shape, as in Figure 8.2, so that the term *volatility smirk* has gained popularity. The 'smiling' shape is often more pronounced for FX options than for equity options.

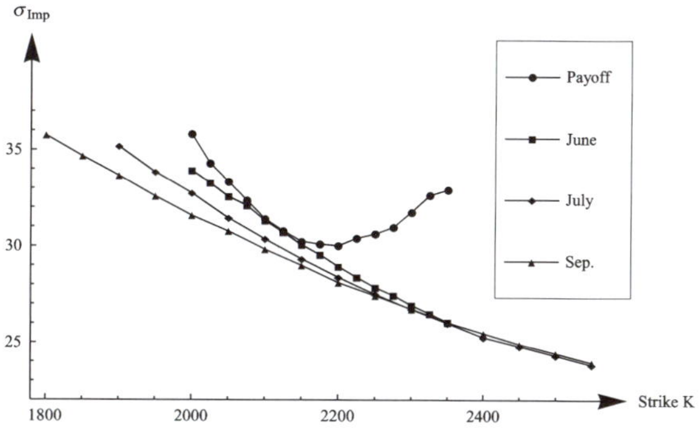

Fig. 8.2 Implied volatility for call options as function of the strike, for four different maturities (dates as in Figure 4.1)

to achieve a balance between computability and explanatory power, and one can understand models as a tool for decision making, rather than an exact description of reality.

However, most major market participants base their pricing and decision making on more complex and flexible market models than the Black-Scholes model. In practice, models that keep the Black-Scholes structure while aiming to overcome some of its deficiencies prove popular, and we will now discuss some such modifications to the Black-Scholes model.

8.2 The Dupire Model

The *Dupire* or *local volatility* model was simultaneously introduced by Bruno Dupire and by Emannel Derman and Iraj Kani in 1994, to offer a model that can explain the volatility smile. The main idea of this model is to write the volatility as a function of time t and stock price S_t. This is different from the classical Black-Scholes model, where the volatility parameter is kept constant. In particular, the stock price is now modeled as an Itô process with the dynamics

$$\frac{dS_t}{S_t} = \mu(S_t, t)dt + \sigma(S_t, t)dW_t. \tag{8.2}$$

One can obtain a solution to this SDE if both μ and σ are bounded and continuously differentiable.[4]

Employing Itô's formula (6.7) and arguments similar to those used to derive the Black-Scholes formula, it can be shown that the price $C(S, t)$ of a European call option in the Dupire model with $S = S_t$ has to satisfy the partial differential equation

[4]These assumptions can be relaxed to a certain extent. For a detailed discussion of stochastic differential equations, see Øksendal [60].

$$\frac{\partial C}{\partial t} + \frac{\sigma^2(S,t)S^2}{2}\frac{\partial^2 C}{\partial S^2} + rS\frac{\partial C}{\partial S} - rC = 0 \tag{8.3}$$

for $S \in (0, \infty)$ with boundary condition $C(S_T, T) = (S_T - K)^+$.

Note that the call price in the Dupire model is independent of the mean log return μ of the underlying, and the only difference with respect to the Black-Scholes formula lies in the non-constant volatility $\sigma(S_t, t)$. If the volatility is independent of the stock price, i.e. $\sigma(S_t, t) = \sigma(t)$, then (8.3) can be solved explicitly and the call price is given by

$$C(S_t, K, t, T) = S_t\Phi(d_{t+}) - e^{-r(T-t)}K\Phi(d_{t-}), \tag{8.4}$$

with

$$d_{t\pm} = \frac{\log(S_t/K) + (T-t)r \pm \frac{1}{2}\int_t^T \sigma(s)^2\,ds}{\sqrt{\int_t^T \sigma(s)^2\,ds}}.$$

The classical Black-Scholes formula is regained by setting $\sigma(t) \equiv \sigma$.

The relationship between volatility and time is also called *term structure of volatility*. However, from (8.4) it follows for the time-dependent Dupire model that the implied volatility terms $\int_t^T \sigma(s)ds$ are constant for options of the same maturity – even for different strikes. Hence, the time-dependent volatility model is also not consistent with market option prices, and thus the volatility should be a function of both the maturity *and* the stock price. Such a model would be able to describe practically any volatility structure, as will be further discussed in Chapter 12. In this general setup, (8.3) can no longer be solved explicitly and one will require numerical or Monte Carlo methods to obtain option prices (see Chapters 10 and 11). Finally, note that the Dupire model is a *complete* market model. Heuristically this means that it is possible to replicate any derivative through continuous trading in the underlying and a risk-free account. The exact replication, however, requires knowledge of the volatility function, which is not available in practical applications.

8.3 The Heston Model

While the Dupire model was able to describe the volatility smile, it did not provide additional insight in the dynamics of the stock price. We will therefore turn to another extension of the Black-Scholes model which admits an economic interpretation and contains some important properties observed in empirical price processes. Assume the volatility of the stock price process is itself a stochastic process $\{\sigma_t : t \geq 0\}$, and consider the local dynamics of the stock price S_t

$$\frac{dS_t}{S_t} = \mu dt + \sigma_t dW_t, \tag{8.5}$$

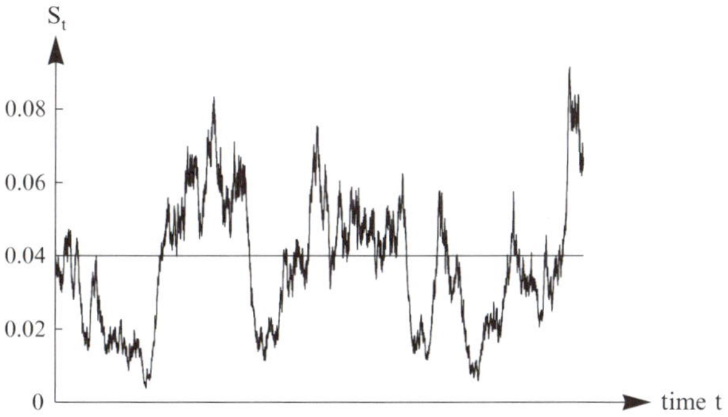

Fig. 8.3 Sample path of a CIR process with $v_0 = 0.04$, $\kappa = 2.5$, $\theta = 0.04$ and $\lambda = 0.3$

where W_t is again a standard Brownian motion. The *stochastic volatility* model that is most popular in practice today was suggested by Steven Heston in 1993, and is commonly referred to as the *Heston model*. In this model the variance of the stock price (i.e. the squared volatility) $v_t = \sigma_t^2$ follows the Itô process

$$dv_t = \kappa(\theta - v_t)dt + \lambda\sqrt{v_t}d\widetilde{W}_t, \tag{8.6}$$

with $\kappa, \theta, \lambda, v_t > 0$ and $\widetilde{W}_t$ being a Brownian motion (the so-called *Cox-Ingersoll-Ross (CIR) process*). This process is *mean-reverting*, which means that its drift term $\kappa(\theta - v_t)$ is (strictly) positive if v_t is smaller than the *long-term mean* θ, and negative if $v_t > \theta$. Therefore, the process v_t keeps drifting towards θ, i.e. it fluctuates around θ based on a diffusion. The parameter κ defines the speed of the mean-reversion, and λ is often called *volatility of volatility* (even though it strictly speaking describes the volatility of the variance). As the Brownian motion $\widetilde{W}_t$ is time-continuous with probability 1, the same holds true for the variance process v_t. Note that $v_t = 0$ gives $dv_t = \kappa\theta dt$ with $\kappa, \theta > 0$, so that with probability 1 the process v_t does not turn negative, which is fundamental for the consistency of the model. If *Feller's condition*

$$2\kappa\theta > \lambda \tag{8.7}$$

holds, the CIR process is even positive with probability 1. Figure 8.3 depicts a sample path of a CIR process.

The Brownian motions W_t and $\widetilde{W}_t$ in (8.5) and (8.6) are assumed to correlate with correlation coefficient ρ $(-1 < \rho < 1)$, meaning that $\mathrm{Cov}(W_t, \widetilde{W}_t) = \rho t$. $\widetilde{W}_t$ can be written as $\widetilde{W}_t = \rho W_t + \sqrt{1-\rho^2}W_t^*$, where W_t^* is a Brownian motion independent of W_t, and it can be shown that (see Exercise 2)

$$dW_t d\widetilde{W}_t = \rho dt. \tag{8.8}$$

Empirical studies suggest that the volatility typically rises in times of falling stock prices and falls when stock prices go up. This effect is often referred to as *leverage effect*, and as a consequence ρ is typically chosen negative. Let us now take a closer look at the calculation of the price of a European call option in the Heston model. As when deriving the Black-Scholes formula, we aim at constructing a risk-free portfolio consisting of a short position in a call option, $\Delta = \Delta_t$ units of the underlying and a cash position of $C(S, v, t) - \Delta S_t$. Then

$$d\pi_t = \Delta_t dS + (C(S, v, t) - \Delta S_t) r dt - dC(S, v, t)$$

are the local dynamics of the portfolio value π_t at time t. As the variance $v = v_t$ of the stock price is a stochastic process itself, C is explicitly written as a function of also v in the above notation. In order to be able to employ similar arguments as for hedging in the Black-Scholes model, the local dynamics of the option price has to be computed. (6.7) only gave Itô's Lemma for one-dimensional Itô processes, however, the more-dimensional version can be easily derived following the same logic.[5] A Taylor series expansion in combination with (6.5) and (8.8) leads to

$$\begin{aligned} dC(S, v, t) = &\left(\frac{\partial C}{\partial S} \mu S + \frac{\partial C}{\partial v} \kappa(\theta - v) + \frac{\partial C}{\partial t} + \frac{1}{2} \frac{\partial^2 C}{\partial S^2} v S^2 + \frac{1}{2} \frac{\partial^2 C}{\partial v^2} \lambda^2 v \right. \\ &\left. + \frac{\partial^2 C}{\partial v \partial S} \rho v \lambda S \right) dt + \frac{\partial C}{\partial S} \sqrt{v} S dW_t + \frac{\partial C}{\partial v} \lambda \sqrt{v} d\widetilde{W}_t \end{aligned} \tag{8.9}$$

and ultimately to

$$\begin{aligned} d\pi_t = &\left(\Delta_t - \frac{\partial C}{\partial S} \right) dS - \frac{\partial C}{\partial v} dv + (C(S, v, t) - \Delta S_t) r dt \\ &- \left(\frac{\partial C}{\partial t} + \frac{1}{2} \frac{\partial^2 C}{\partial S^2} v S^2 + \frac{1}{2} \frac{\partial^2 C}{\partial v^2} \lambda^2 v + \frac{\partial^2 C}{\partial v \partial S} \rho v \lambda S \right) dt \end{aligned}$$

It is easy to see that setting $\Delta = \partial C / \partial S$ eliminates the dependence of the stock price on the change *dS*. On the other hand, the stochastic component of the variance $v = v_t$ does not disappear, so that the price of a European call option in the Heston model cannot be determined uniquely. It is concluded that not all derivative pay-offs can be replicated by a portfolio consisting of only the underlying and the risk-free account. We now face a whole range of possible call prices that are in accordance with no-arbitrage requirements (which is a general property of so-called *incomplete* market models). With the exception of the binomial, the Black-Scholes and the Dupire model, all other market models used in practice are incomplete.

For the Heston model, adding a simple condition will allow to find the unique price of the derivative. One adds $-\Lambda$ units of an additional hedging instrument (here:

[5] **Multi-dimensional Itô's Lemma**. Let $\mathbf{X}_t = (X_{t,1}, X_{t,2}, ..., X_{t,n})$ be an n-dimensional vector of Itô processes. For twice differentiable f it then follows that $df(t, \mathbf{X}_t) = \frac{\partial f}{\partial t} dt + \sum_{i=1}^{n} \frac{\partial f}{\partial X_{t,i}} dX_{t,i} + \frac{1}{2} \sum_{1 \leq i,j \leq n} \frac{\partial^2}{\partial X_{t,i} \partial X_{t,j}} f \cdot dX_{t,i} dX_{t,j}$.

an otherwise identical call option $\widetilde{C}$ with a longer maturity) to the portfolio, and the local dynamics

$$d\pi_t = \Delta_t dS + (C(S,v,t) - \Delta S_t - \Lambda\tilde{C}(S,v,t))r dt - dC(S,v,t) + \Lambda\, d\tilde{C}(S,v,t)$$

leads to

$$\begin{aligned} d\tilde{\pi}_t = &\ (C(S,v,t) - \Delta S_t - \Lambda\tilde{C}(S,v,t))r dt \\ &- \left(\frac{\partial C}{\partial t} + \frac{1}{2}\frac{\partial^2 C}{\partial S^2} v S^2 + \frac{1}{2}\frac{\partial^2 C}{\partial v^2}\lambda^2 v + \frac{\partial^2 C}{\partial v \partial S}\rho v \lambda S\right) dt \\ &+ \Lambda\left(\frac{\partial \widetilde{C}}{\partial t} + \frac{1}{2}\frac{\partial^2 \widetilde{C}}{\partial S^2} v S^2 + \frac{1}{2}\frac{\partial^2 \widetilde{C}}{\partial v^2}\lambda^2 v + \frac{\partial^2 \widetilde{C}}{\partial v \partial S}\rho v \lambda S\right) dt \\ &+ \left(\Lambda\frac{\partial \widetilde{C}}{\partial S} - \frac{\partial C}{\partial S} + \Delta\right) dS + \left(\Lambda\frac{\partial \widetilde{C}}{\partial v} - \frac{\partial C}{\partial v}\right) dv. \end{aligned}$$

Choosing[6]

$$\Lambda = \frac{\partial C/\partial v}{\partial \widetilde{C}/\partial v} \quad \text{and} \quad \Delta = \frac{\partial C}{\partial S} - \Lambda\frac{\partial \widetilde{C}}{\partial S}$$

eliminates the stochastic component in the dynamics $d\tilde{\pi}_t$ and makes the portfolio $\tilde{\pi}$ risk-free. Hence the deterministic drift term has to vanish as well, which leads to the condition

$$\frac{\dfrac{\partial C}{\partial t} + \dfrac{1}{2}\dfrac{\partial^2 C}{\partial S^2} v S^2 + \dfrac{1}{2}\dfrac{\partial^2 C}{\partial v^2}\lambda^2 v + \dfrac{\partial^2 C}{\partial v \partial S}\rho v \lambda S + r\dfrac{\partial C}{\partial S} - rC}{\dfrac{\partial C}{\partial v}} =$$

$$\frac{\dfrac{\partial \widetilde{C}}{\partial t} + \dfrac{1}{2}\dfrac{\partial^2 \widetilde{C}}{\partial S^2} v S^2 + \dfrac{1}{2}\dfrac{\partial^2 \widetilde{C}}{\partial v^2}\lambda^2 v + \dfrac{\partial^2 \widetilde{C}}{\partial v \partial S}\rho v \lambda S + r\dfrac{\partial \widetilde{C}}{\partial S} - r\widetilde{C}}{\dfrac{\partial \widetilde{C}}{\partial v}}.$$

We observe that the above quotient does not depend on the strikes or maturities of the two calls and can hence be written as some function $f(S,v,t)$. For C, as well as for $\widetilde{C}$, we find

[6]Note that Λ is the quotient of the Vegas of the two calls.

$$r(C - \frac{\partial C}{\partial S} S) + f(S, v, t)\frac{\partial C}{\partial v} = \frac{\partial C}{\partial t} + \frac{1}{2}\frac{\partial^2 C}{\partial S^2} v\, S^2 + \frac{1}{2}\frac{\partial^2 C}{\partial v^2}\lambda^2 v + \frac{\partial^2 C}{\partial v \partial S}\rho\lambda S\, v. \tag{8.10}$$

Substituting the above into equation (8.9) yields

$$dC(S, v, t) = \left(rC + (\mu - r)\frac{\partial C}{\partial S} S + \big(f(S, v, t) + \kappa(\theta - v)\big)\frac{\partial C}{\partial v} \right) dt$$
$$+ \frac{\partial C}{\partial S}\sqrt{v} S dW_t + \frac{\partial C}{\partial v}\lambda \sqrt{v}\, d\widetilde{W}_t.$$

This indicates that the difference in the *expected* returns of (a) an investment in the call, versus (b) an investment of C in the risk-free account is given by

$$(\mu - r)\frac{\partial C}{\partial S} S + \big(f(S, v, t) + \kappa(\theta - v)\big)\frac{\partial C}{\partial v}.$$

The first term can be interpreted as a compensation for the risk of the share price changing (see Exercise 2 in Chapter 14) as seen in the Black-Scholes model, while the second term compensates for volatility risk. The quantity $f(S, v, t) + \kappa(\theta - v)$ is typically referred to as *market price of volatility risk*, and Heston assumes it to be of the form $\nu \cdot v$.[7]

(8.10) must hold for all $S, v \geq 0$, so that this choice then leads to

$$\frac{\partial C}{\partial S} rS + \frac{\partial C}{\partial t} + \frac{1}{2}\frac{\partial^2 C}{\partial S^2} vS^2 + \frac{1}{2}\frac{\partial^2 C}{\partial v^2}\lambda^2 v + \frac{\partial^2 C}{\partial v \partial S}\rho\lambda S\, v \tag{8.11}$$
$$+ \frac{\partial C}{\partial v}(\kappa\theta - (\kappa + \nu)v) - rC = 0,$$

for $S \in (0, \infty)$ and $v \in (0, \infty)$ and with suitable boundary conditions. In general, this equation can be solved numerically, but not explicitly. A method that is often computationally faster than solving (8.11) numerically can be motivated as follows. Based on the Fundamental Theorem of Asset Pricing the price of the call can be written as its discounted expected pay-off under the risk-neutral probability measure. Defining the market price of volatility determines the risk-neutral measure, and it follows that

$$C(S, v, t) = \mathbb{E}^{\mathbb{Q}}\left[e^{-r(T-t)}(S_T - K)^+\right], \tag{8.12}$$

with $S = S_t$ and $v = v_t$ (under $\mathbb{Q}$) satisfying the differential equations

[7] This assumption has been much debated, but has the advantage that the resulting model remains analytically tractable.

$$\frac{dS_t}{S_t} = rdt + \sqrt{v_t}\, dW_t^{\mathbb{Q}}, \qquad (8.13)$$
$$dv_t = \tilde{\kappa}(\tilde{\theta} - v_t)dt + \lambda\sqrt{v_t}\, d\widetilde{W}_t^{\mathbb{Q}},$$

with $W^{\mathbb{Q}}$ and $\widetilde{W}^{\mathbb{Q}}$ being two correlated Brownian motions under $\mathbb{Q}$ with correlation coefficient ρ, and $\tilde{\kappa} = \kappa + \nu$ and $\tilde{\theta} = \kappa\theta/(\kappa + \nu)$. Compared to (8.6) only the parameters κ, θ have changed, and the drift term in (8.5) has been modified.[8]

The evaluation of formula (8.12) requires the knowledge of the distribution of $\log(S_t)$ based on (8.13). This distribution can be defined by its *characteristic function*:[9]

$$\begin{aligned}\mathbb{E}[e^{iu\log(S_T)}] &= \exp(iu(\log S_0 + rT)) \qquad (8.14)\\ &\times \exp(\tilde{\theta}\tilde{\kappa}\lambda^{-2}((\tilde{\kappa} - \rho\lambda iu - d)T - 2\log((1 - ge^{-dT})/(1 - g))))\\ &\times \exp(\sigma_0^2\lambda^{-2}(\tilde{\kappa} - \rho\lambda iu - d)(1 - e^{-dT})/(1 - ge^{-dT})),\end{aligned}$$

with

$$d = \sqrt{(\rho\lambda ui - \tilde{\kappa})^2 + \lambda^2(iu + u^2)},$$
$$g = (\tilde{\kappa} - \rho\lambda iu - d)/(\tilde{\kappa} - \rho\lambda iu + d).$$

Once the characteristic function is known, efficient algorithms exists for solving (8.12) numerically (see Chapter 12).

Note that equations (8.11) and (8.12) give a no-arbitrage call price for *arbitrary* choices of ν. This means, in particular, that the call price can no longer be determined uniquely (see Exercise 8). Conversely, the parameters $\tilde{\kappa}$ and $\tilde{\theta}$, which are influenced by ν, can be directly obtained from the market prices of traded options (see Chapter 12).

8.4 Price Jumps and the Merton Model

None of the previously discussed models has considered 'jumps', i.e. points of discontinuity of the stock price process. However, even when excluding small unexceptional jumps that could potentially be approximated by a diffusion, jumps are actually observed in the stock markets. To state an extreme example of a market

[8] Formula (8.12) with (8.13) can be directly deduced through the principle of risk-neutral valuation, without considering the partial differential equation. The derivation of this, however, is beyond the scope of this book.

[9] The characteristic function $\mathbb{E}(e^{iuX})$ of a random variable X can give a very efficient way to describe the properties of the distribution of X. If X has the probability density function f_X, then $\mathbb{E}(e^{iuX})$ can be interpreted as the Fourier transform of f_X.

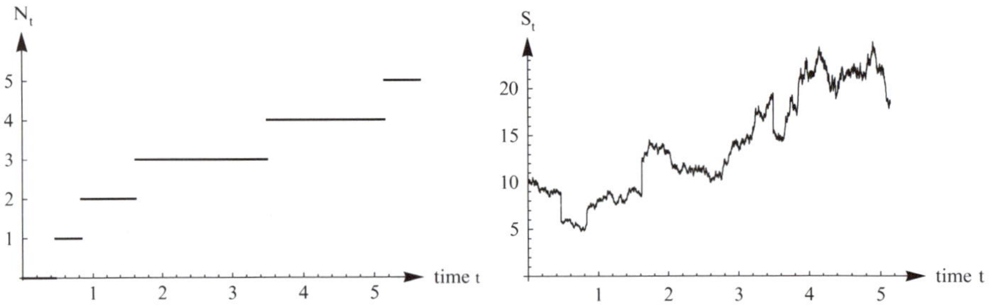

Fig. 8.4 Sample path of a Poisson process (left) and of a corresponding sample path of the stock price process in the Merton model (right)

jump, on the 'Black Monday' (19th October 1987) the Dow Jones index fell as much as 22.5 % within one single trading day. We shall now define a stochastic process that will be an important tool for incorporating jumps in stochastic models.

Definition. Let $\{X_i, i \in \mathbb{N}\}$ be independent and exponentially distributed inter-jump times with parameter λ, i.e.

$$\mathbb{P}[X_i \leq x] = 1 - e^{-\lambda x} \qquad \text{for all } i \in \mathbb{N}.$$

Furthermore, let $\tau_n = \sum_{i=1}^{n} X_i$ be the time of the n-th jump. Then,

$$N_t = \sum_{i \geq 1} 1_{\{\tau_i \leq t\}} \tag{8.15}$$

is a *Poisson process* with *intensity* λ.[10]

The process N_t (cf. Figure 8.4) counts the number of jumps up to time t (e.g. jumps due to market crashes or rumors on potential take-overs), and the time intervals between two such jumps are Exp(λ) distributed. Let Y_i be the size of the i-th jump, with the $(Y_i)_{i\geq 1}$'s being independent and identically distributed. The process

$$Z_t = \sum_{i=1}^{N_t} Y_i \tag{8.16}$$

is then called *compound Poisson process*. As early as in 1976, this process class was used by R. Merton for the modeling of stock prices. The so-called *Merton model*

[10]The above is only one possible way of defining a Poisson process. The name is based on the fact that N_t follows a Poisson distribution. This distribution was introduced by the French physicist and mathematician Siméon-Denis Poisson (1781–1840) in his 1837 work 'Research on the Probability of Judgments in Criminal and Civil Matters'.

(a so-called *jump diffusion model*) describes the local dynamics of the stock price S_t (see Figure 8.4) as

$$\frac{dS_t}{S_t} = \mu dt + \sigma dW_t + dZ_t, \tag{8.17}$$

with a Brownian motion W_t that is independent of Z_t, constant volatility σ and log-normally distributed jump sizes.

In contrast to the Heston model, a call cannot be hedged by another call in the Merton model.[11] However, the price of a European call option can once again be written as its discounted expected pay-off under the risk neutral measure $\mathbb{Q}$, i.e.

$$C(S_t, t) = \mathbb{E}^{\mathbb{Q}}[e^{-rT}(S_T - K)^+],$$

with

$$\frac{dS_t}{S_t} = (r - \tilde{\lambda}(e^{\tilde{m}+\tilde{\delta}^2/2} - 1))dt + \sigma dW_t^{\mathbb{Q}} + dZ_t^{\mathbb{Q}}, \tag{8.18}$$

where $Z_t^{\mathbb{Q}}$ is again a compound Poisson process, but with intensity $\tilde{\lambda}$ and jump size parameters $\tilde{m}$ and $\tilde{\delta}$. These three parameters can be freely chosen.[12] Adding the jump component dZ_t has turned the Black-Scholes model into an incomplete market model. The characteristic function of $\log(S_T)$ is here given by $\mathbb{E}^{\mathbb{Q}}[e^{iu\log(S_T)}] = e^{t\psi(u)}$, with

$$\psi(u) = iu\left(r - \frac{\sigma^2}{2} - \tilde{\lambda}(e^{\tilde{m}+\tilde{\delta}^2/2} - 1)\right) - \frac{\sigma^2 u^2}{2} + \tilde{\lambda}\left(e^{iu\tilde{m}-\tilde{\delta}^2 u^2/2} - 1\right).$$

The Merton model is only one of many jump models that have been studied in detail over recent years. A popular class of jump models is given by *Lévy models* whose jump structure is defined by a general Lévy process (i.e. a stochastic process with independent and stationary increments). It is then no longer necessary that the resulting log returns are normally distributed, which better reflects empirical observations (cf. Figure 8.1). For many such models it is still possible to derive an explicit form of the characteristic function of S_t under $\mathbb{Q}$, and efficient numerical algorithms exist to calibrate these models to market data, and to price derivatives (see Chapter 12).

[11] Infinitely many derivatives would be required for hedging in the Merton model.

[12] There exist other risk-neutral measures which 'fit' the Merton model. Mostly, however, one would directly model the risk-neutral process according to (8.18).

8.5 Key Takeaways, References and Exercises

Key Takeaways

After working through this chapter you should understand and be able to explain the following terms and concepts:

- Shortcomings of the Black-Scholes model: empirical skewness/kurtosis, the volatility smile
- The Dupire model: models local volatility $\sigma = \sigma(t, S_t)$, explicit solution for the special case $\sigma = \sigma(t)$
- The Heston model: models variance σ_t^2 as a mean-reverting Itô process, no explicit solution, a differential equation can be solved numerically to give European option prices
- The Merton model: uses Black-Scholes setup and adds jump (compound Poisson) process when modeling dS_t/S_t, numerical solution possible

References

The local volatility model was introduced by Dupire [24] and Derman & Kani [21]. The Heston model was first published in [40]. For a more detailed discussion around the here presented and other volatility models, the interested reader is referred to the books Fouque, Papanicolaou & Sircar [35] and Lewis [52]. A comprehensive discussion of the modeling with Lévy processes is provided in Schoutens [69] and Cont & Tankov [18].

Exercises

1. Show that equation (8.1) has a *unique* solution for all market prices that satisfy the trivial no-arbitrage inequalities (4.2).
2. Prove that $\mathbb{E}\big[\big(\Delta W_t \Delta \widetilde{W}_t/(\rho\Delta t)\big)^2\big] \to 1$ for $\Delta t \to 0$, with $\rho(t) = \mathrm{Cov}[W_t, \widetilde{W}_t]$.
3. Show that the exponential distribution fulfills the 'no-memory' property, i.e. that for $X \sim \mathrm{Exp}(\lambda)$ it holds that $\mathbb{P}[X > x + y | X > x] = \mathbb{P}[X > y]$ for $y > 0$.
4. Let $Z = \sum_{i=1}^{N_t} Y_i$ be a compound Poisson process with intensity parameter λ and the characteristic function of the jump sizes shall be given by ϕ_Y. Prove that

$$\mathbb{E}[\exp(\mathrm{i}z \sum_{i=1}^{N_t} Y_i)] = \exp\left(\lambda\, t (\phi_Y(z) - 1)\right).$$

   (Hint: start by conditioning on the number of jumps N_t and use that the Y_i's are independent (also of N_t) and identically distributed.)
5. Compute $d\log(S_t)$ in the Heston model for S_t via (8.13).
6. An extension to Itô's formula for jump processes uses (8.18) to yield

$$d\log(S_t) = (r - \sigma^2/2 - \tilde{\lambda}(e^{\tilde{m}+\tilde{\delta}^2/2} - 1))dt + \sigma dW_t^{\mathbb{Q}} + d\tilde{Z}_t^{\mathbb{Q}}$$

and, hence,

$$\log(S_t) = (r - \sigma^2/2 - \widetilde{\lambda}(e^{\widetilde{m}+\widetilde{\delta}^2/2} - 1))t + \sigma W_t^{\mathbb{Q}} + \widetilde{Z}_t^{\mathbb{Q}}.$$

Hereby, $\widetilde{Z}_t^{\mathbb{Q}}$ is a compound Poisson process with the same intensity $\widetilde{\lambda}$ as $Z_t^{\mathbb{Q}}$ in (8.18). In particular, we have

$$\widetilde{Z}_t^{\mathbb{Q}} = \sum_{i=1}^{N_t^{\mathbb{Q}}} Y_i,$$

where $N_t^{\mathbb{Q}}$ is a Poisson process with intensity $\widetilde{\lambda}$ and the Y_i's are normally distributed with mean $\widetilde{m}$ and variance $\widetilde{\delta}$. Use the above formula to derive the characteristic function of the stock log price in the Merton model. (Hint: Use that the Brownian motion and the compound Poisson process are stochastically independent.)

Interest Rate Models

9

So far we have assumed that interest rates are given either as constants or as deterministic functions of time. However, in reality interest rates show stochastic behavior (cf. Fig. 1.2). While this often only plays a secondary role when dealing with stock derivatives, it is, of course, the core aspect when pricing interest rate derivatives. After a brief introduction to some of the most commonly traded interest rate products, this chapter will present a selection of popular interest rate models.

9.1 Caps, Floors and Swaptions

In Chapters 1 and 2 we have encountered interest rate swaps and bonds as examples of interest-dependent products. Bonds are sometimes traded at exchanges, while most other interest rate products are traded exclusively OTC. Even though traded OTC only, standard interest-rate products are typically very liquid. We shall now discuss some of these standard products in more detail.

Caps and Floors

Let us start with an example. Assume that a borrower pays floating interest (e.g. semi-annual payments of Euribor6M+1.25%) on some outstanding loan principal (e.g. 10,000 EUR).[1] In order to lower the risk of the borrower not being able to make interest payments in full if Euribor6M rises, it is agreed that the interest rate shall not exceed 6% (one says: the interest rate is *capped* at 6%).

Interest is typically paid *in arrears* so that the interest payment at time t_i is determined by the rate at time t_{i-1}. More precisely, the interest payment at time t_i is given as

[1]The principal may decrease over time when the borrower amortizes the loan.

H. Albrecher et al., *Introduction to Quantitative Methods for Financial Markets*, Compact Textbooks in Mathematics, DOI 10.1007/978-3-0348-0519-3_9,

$$\min(\text{Euribor6M}(t_{i-1}) + 1.25\,\%, 6\,\%)$$
$$= 1.25\,\% + \text{Euribor6M}(t_{i-1}) - (\text{Euribor6M}(t_{i-1}) - 4.75\,\%)^+.$$

A (European) pay-off $(R - K)^+$, where R is the reference interest rate (e.g. Euribor6M), is called *caplet* with strike price K, and is structurally similar to a European call on a stock. Interest payments for fixed income products are made on scheduled dates $(t_1, \ldots, t_n)$, and collecting caplets for these single payment dates into one single contract gives a so-called *cap*. A cap has the discounted pay-off

$$\sum_{i=1}^{n} D(t_i)(t_i - t_{i-1})(R(t_{i-1}) - K)^+,$$

where $D(t)$ is the discount factor today for payments made at time t. For a plain vanilla cap, the reference interest rate R applied to the period $(t_{i-1}, t_i]$ is determined at t_{i-1}, while the corresponding payment is made at t_i.

Similarly, changing the above pay-off to $(K - R)^+$ gives a *floorlet*, and single floorlets can again be collected to give a *floor*.[2]

Swaptions

A *swaption* is an option to enter into a pre-specified swap contract at option expiry T. One distinguishes *payer* and *receiver* swaptions, depending on what type of swap the swaption buyer can enter. For example, a 3×8 payer swaption with strike price 4.5 % gives the right (but not the obligation) to become in $T = 3$ years (the maturity of the European option) the fixed rate payer (at 4.5 % per year) in an 8-year interest rate swap. If the 8-year swap rate lies above 4.5 % in 3 years, this payer swaption will be exercised. If the 8-year swap rate turns out below 4.5 %, the option to enter the 4.5 % swap expires unused. Note that the swaption counterparties might simply agree to *cash-settle* the contract at option expiry T according to the market value of the underlying swap then, rather than physically entering into the swap contract. The discounted pay-off of a payer swaption is given by

$$\left(\sum_{i=1}^{n} D(t_i)(t_i - t_{i-1})(\text{sr}_T - K)\right)^+,$$

where $T = t_0$ is the expiry of the swaption, $(t_1, t_2, \ldots, t_n)$ are the payment dates of the underlying swap, sr_T is the applicable market swap rate at time T and the $D(t_i)$'s are appropriate discount factors. The most-quoted swaptions are *at-the-money swaptions*.[3] The strike rate K is hereby set at the forward swap rate (in the

[2] The terms cap and floor are based on the fact that the contracts allow to define upper and lower bounds for future interest payments, as illustrated in the above example.

[3] An option is called 'at-the-money' if the current price of the underlying equals the strike price of the option contract. Similarly, one refers to *in-the-money* and *out-of-the-money* if the option would

above example, this would be the fixed rate for an 8-year swap that *becomes effective* in 3 years from now, such that this swap has a fair value of 0 today). Option expiries range from a few months to 30 years, and swap terms of 1 up to 50 years.

9.2 Short-Rate Models

In Section 1.3 we described the local dynamics of a bank account B_t at time t as

$$dB_t = B_t r dt,$$

which led to $B_t = B_0 e^{rt}$ for an initial account balance of B_0 and some constant interest rate r. *Short-rate models* generalize this by modeling the interest rate itself as a stochastic process, i.e. $r = r_t$, which yields

$$dB_t = B_t r_t dt \quad \text{and, respectively,} \quad B_t = B_0 e^{\int_0^t r_s ds}.$$

The *short rate*[4] r_t is typically modeled as an Itô process

$$dr_t = g(r_t, t)dt + h(r_t, t)dW_t, \tag{9.1}$$

where W_t is a Brownian motion and g and h are to-be-determined functions.

In such a setup, one must now also use the stochastic interest rate when discounting future cash flows. The fair value of many interest rate products (such as bonds, caps or swaptions) is then also driven by the discount factors $e^{-\int_0^t r_s ds}$ (and hence by the process r_t). Assume that the fair value V of an interest rate product (which may include optional features) depends, apart from its contract specifications, only on current time t and rate r_t, i.e. $V = V(r_t, t)$. We now aim to describe the dynamics dV under the model (9.1).

Recall the Black-Scholes model, in which it was possible to compile a risk-neutral portfolio from a short position in a call option, Δ stocks and cash. In this context it was essential to have a tradeable underlying stock. In the case of short rates, the underlying is a theoretical structure rather than a traded financial product.

Nevertheless, similar to hedging in the Heston model (see Section 8.3), this issue can be overcome by constructing a risk-neutral portfolio π_t that consists of a short position in the original interest rate product with maturity T_1 and price V_1, Δ units in another product with different maturity T_2 and price V_2 and a position of $V_1 - \Delta V_2$ in the risk-less asset. Itô's Lemma (6.7) then gives (for sufficiently regular parameter functions g and h) the dynamics $d\pi_t$ as (using the notation $r = r_t$)

have positive pay-out or no pay-out, respectively, if it was exercised immediately rather than at maturity. In this context, strike prices are often quoted in terms of a percentage of the current price of the underlying.

[4]The term short can be understood from the interpretation $r_t = r(t; t + dt)$ (cf. Section 1.5).

$$d\pi_t = (V_1 - \Delta V_2) r_t dt - \left(\frac{\partial V_1}{\partial r} g(r,t) + \frac{\partial V_1}{\partial t} + \frac{1}{2} h^2(r,t) \frac{\partial^2 V_1}{\partial r^2} \right) dt - \frac{\partial V_1}{\partial r} h(r,t) dW_t$$
$$+ \Delta \left[\left(\frac{\partial V_2}{\partial r} g(r,t) + \frac{\partial V_2}{\partial t} + \frac{1}{2} h^2(r,t) \frac{\partial^2 V_2}{\partial r^2} \right) dt + \frac{\partial V_2}{\partial r} h(r,t) dW_t \right].$$

Choosing $\Delta := \frac{\partial V_1}{\partial r} / \frac{\partial V_2}{\partial r}$ allows to eliminate the stochastic terms in the above equation. In order to rule out arbitrage opportunities, also the deterministic drift needs to equal 0 and thus leads to

$$\left(\frac{\partial V_1}{\partial t} + \frac{1}{2} h^2(r,t) \frac{\partial^2 V_1}{\partial r^2} - \left(\frac{\partial V_1}{\partial r} \Big/ \frac{\partial V_2}{\partial r} \right) \cdot \left(\frac{\partial V_2}{\partial t} + \frac{1}{2} h^2(r,t) \frac{\partial^2 V_2}{\partial r^2} \right) \right) dt$$
$$= r \left(V_1 - V_2 \cdot \frac{\partial V_1}{\partial r} \Big/ \frac{\partial V_2}{\partial r} \right) dt.$$

Rearranging this equation yields

$$\frac{\frac{\partial V_1}{\partial t} + \frac{1}{2} h^2(r,t) \frac{\partial^2 V_1}{\partial r^2} - r V_1}{\frac{\partial V_1}{\partial r}} = \frac{\frac{\partial V_2}{\partial t} + \frac{1}{2} h^2(r,t) \frac{\partial^2 V_2}{\partial r^2} - r V_2}{\frac{\partial V_2}{\partial r}}. \tag{9.2}$$

Note that V_1 and V_2 as above depend on t, r and on the respective contractual specifics. Thus Equation (9.2) can hold for *all* r, only if both sides of the equation just depend on r and t, but not on the characteristics of V_1 or V_2. Denoting either of the quotients in (9.2) by $\omega(r,t)$, we conclude that the fair value V of any derivative depending on r solves

$$\frac{\partial V}{\partial t} + \frac{1}{2} h^2(r,t) \frac{\partial^2 V}{\partial r^2} - \omega(r,t) \frac{\partial V}{\partial r} - rV = 0 \qquad \text{for } r \in \mathbb{R},$$

where $\omega(r,t)$ corresponds to the function f in the Heston model and is determined by the market price of short-rate risk.

9.3 The Hull-White Model: a Short-Rate Model

Similarly to Section 8.3, the market price of risk will determine the short-rate process under the risk-neutral measure. In particular, $-\omega(r_t,t)$ corresponds to the drift of the short rate process (cf. f in the Heston model). In the Hull-White model (1990), the dynamics of r_t under the risk neutral measure[5] are given as

$$dr_t = \big(a(t) - b(t) r_t\big) dt + \sigma(t) dW_t. \tag{9.3}$$

[5] Throughout the remaining chapter we will omit explicitly writing $\mathbb{Q}$ to keep the notation compact.

Note that this process is mean-reverting for positive $b(t)$ (cf. CIR processes in Chapter 8), so that a high short rate will be pulled downwards and a low rate upwards. Particular cases of this model are the *Vasiček model* (1977), in which the parameter functions $a(t)$, $b(t)$ and $\sigma(t)$ are constants, or the *Ho-Lee model*, in which $b(t) \equiv 0$. The Vasiček model is the easiest one to handle analytically, however, it typically does not provide satisfactory results when fitted to yield curves implied by observed market prices. The Ho-Lee model, on the other hand, has the shortcoming that the short rate is not mean-reverting, and it crosses any upper (lower) bound with probability 1 at some point in time if $a(t) \geq 0$ ($a(t) \leq 0$).[6] Throughout the remaining part of this section we will assume $b(t) \equiv b$ and $\sigma(t) \equiv \sigma$ to be constant, see also Exercise 6 (the general case can be analyzed with the same methods, but the notation will be more cumbersome).[7]

In contrast to the Black-Scholes model, σ in (9.3) is not multiplied by the value process (here: the short rate r_t), so that σ can be interpreted as the volatility of r_t for low *speeds of reversion* b. Also note that the short rate has a positive probability of becoming negative (which is often seen as a disadvantage of the model).

For $V = V(r_t, t)$[8], the considerations in the previous section lead to the *Hull-White differential equation*

$$\frac{\partial V}{\partial t} + \frac{1}{2}\sigma^2 \frac{\partial^2 V}{\partial r^2} + \big(a(t) - b\,r\big)\frac{\partial V}{\partial r} - rV = 0 \qquad (r \in \mathbb{R}), \tag{9.4}$$

where we have used that $-\omega(r,t) = a(t) - b\,r$. The individual characteristics of the respective instruments will again be mirrored in the boundary and possible jump conditions (for example, for times where coupon payments are made).

Let us now study in more detail the case of a zero-coupon bond with principal 1 and maturity T. We aim to derive the price $Z(r_t, t; T) := V(r_t, t)$ of this bond today (at $t = 0$), and we assume to know the current short rate r_0. To find the price, one must solve the PDE (9.4) with the boundary condition

$$Z(r, T; T) = 1.$$

Since (9.4) is linear in V, we 'guess' that the price Z at time t is of the form

$$Z(r, t; T) = \exp\big(\alpha(t; T) + r\beta(t; T)\big). \tag{9.5}$$

Substituting the above into (9.4) yields

[6] In the literature the model (9.3) is often referred to as *extended Vasiček model* and the term Hull-White model is often only used for the case where $b(t) \equiv b$ and $\sigma(T) \equiv \sigma$ are constants.

[7] In practice, $b(t)$ typically shows low variability, while $\sigma(t)$ can depend significantly on time.

[8] In Chapter 13 we will consider the example of a *snowball*, where the dependence structure will be more complicated.

$$0 = r \cdot Z(r,t;T)\Big(\beta'(t;T) - b\beta(t;T) - 1\Big)$$
$$+Z(r,t;T)\left(\alpha'(t;T) + \frac{1}{2}\sigma^2\beta^2(t;T) + a(t)\beta(t;T)\right)$$

for arbitrary values r. As the price of a zero-coupon bond must be positive, we arrive at a system of two ODEs,

$$\beta'(t;T) - b\beta(t;T) - 1 = 0, \quad \alpha'(t;T) + \frac{1}{2}\sigma^2\beta^2(t;T) + a(t)\beta(t;T) = 0, \tag{9.6}$$

with the boundary conditions $\alpha(T;T) = 0$ and $\beta(T;T) = 0$. It is straightforward to solve this system (see Exercise 3) and one obtains the solution

$$\begin{aligned} \beta(t;T) &= -\frac{1}{b}\left(1 - e^{-b(T-t)}\right), \\ \alpha(t;T) &= \frac{\sigma^2}{2b^2}\left(\beta(t;T) - \frac{1}{2}\beta(t;T)^2 - (T-t)\right) - \int_t^T a(s)\beta(s;T)\,ds. \end{aligned} \tag{9.7}$$

To summarize, the Hull-White model determines the fair price at time $t = 0$ of a zero-coupon bond with principal 1 and maturity T and with initial short rate r_0 as

$$Z(r_0, 0; T) = \exp\left(\alpha(0;T) + r_0\beta(0;T)\right). \tag{9.8}$$

More generally, $Z(r_t, t; T)$ at time $t < T$ is given by (9.5) with $\alpha(t;T)$ and $\beta(t;T)$ given by (9.7), and $r = r_t$ (r_t is unknown at time $t = 0$, so the time t price of the zero-coupon bond is a random variable).

Consider now a European call option with expiry T and strike price K on a zero-coupon bond with maturity $T + S$ ($S > 0$). The pay-off of this option, whose value at time t is denoted by $\mathrm{ZC}(r_t, t; T, S, K)$, is

$$\big(Z(r_T, T; T+S) - K\big)^+ = \big(\exp\left(\alpha(T, T+S) + r_T\beta(T, T+S)\right) - K\big)^+.$$

Since this pay-off depends only on r_T, ZC must again satisfy (9.4) with the boundary condition

$$\mathrm{ZC}(r, T; T, S, K) = \big(\exp\left(\alpha(T, T+S) + r\beta(T, T+S)\right) - K\big)^+$$

and α and β as in (9.7). This equation can be solved explicitly, so that today's price of the call option is

$$\mathrm{ZC}(r_0, 0; T, S, K) = Z(r_0, 0; T+S)\Phi(h_+) - KZ(r_0, 0, T)\Phi(h_-), \tag{9.9}$$

with

$$h_{\pm} = \frac{\log\left(Z(r_0, 0; T + S)/(KZ(r_0, 0, T)) \pm \tilde{\sigma}^2/2\right.}{2\tilde{\sigma}}, \quad \tilde{\sigma}^2 = \frac{1 - e^{-2bT}}{2b}\beta(t; T)^2.$$

Following the above procedure, the price $\mathrm{ZP}(r_0, 0; T, S, K)$ of a put option with pay-off $(K - Z(r_T, T; T + S))^+$ can be determined as

$$\mathrm{ZP}(r_0, 0; T, S, K) = KZ(r_0, 0, T)\Phi(-h_+) - Z(r_0, 0; T + S)\Phi(-h_+). \quad (9.10)$$

Note that the prices of call and put options in the Hull-White model are of similar form as those for stock options in the Black-Scholes model. (9.9) also allows the pricing of caps and floors. Let $R_T(T, T + S)$ be the variable reference interest rate (also: floating rate) at time T for the time interval $[T, T + S]$. Since

$$\big(1 + S \cdot R_T(T, T + S)\big)\, Z(r_T, T; T + S) = 1, \quad \text{or}$$
$$S \cdot R_T(T, T + S) = 1/Z(r_T, T; T + S) - 1$$

it can be shown by no-arbitrage arguments (see Exercise 4) that the price C of a caplet on $R_T(T, T + S)$ with strike price K is given by

$$C = S \cdot (1 + K) \cdot \mathrm{ZP}\big(r_0, 0; T, 1/(1 + KS)\big). \quad (9.11)$$

Analogously one can derive a similar formula for floorlets, so that (9.9) and (9.10) can be used to determine prices of caps and floors in the Hull-White model.

Other Short-Rate Models and Common Criticism of One-Factor Models

A main point of criticism of the Hull-White model is the resulting normal distribution of the short rate, and, the implied positive probability of observing negative interest rates. Models that do not have this issue include the *Cox-Ingersoll-Ross (CIR)*, the *Black-Derman-Toy* and the *Black-Karasinski models*. In the *Black-Karasinski model* the logarithm of the short rate $\log(r_t)$ follows the Hull-White dynamics, i.e.

$$d(\log r_t) = (\eta(t) - c\log(r_t))dt + \sigma dW_t.$$

Short rates in this model remain positive and are log-normally distributed. A major drawback of the Black-Karasinski model is, however, that no explicit formulas are available for the pricing of bonds, bond options, caps or swaptions, so that one always has to turn to efficient numerical solution methods.

In the extended CIR model, the short rate r_t follows the differential equation

$$dr_t = (a(t) - b(t)r_t)dt + \sigma(t)\sqrt{r_t}dW_t.$$

This model yields explicit solutions for the prices of bonds, caps or swaptions, provided that all parameters are assumed constant (which is often seen as too strict an assumption for the fitting to market yield curves).

Since $\beta(t, T)$ in (9.7) is always negative, the bond price formula in the Hull-White model shows that an increase in the short rate r_t leads to a decrease in the zero-coupon bond prices for all terms T (and an increase in the corresponding bond yields). Therefore, interest rates $r(t; T)$ for all terms $T - t$ are perfectly correlated at time t, and a *rotation* of the yield curve is not possible. This is also the case in the CIR and the Black-Karasinski model. Models that overcome this weakness include the so-called *market models*, which we will discuss in the next section, and two- or multi-factor models, such as the *two-factor Hull-White model*,

$$\begin{aligned} dr &= (\theta(t) + u - a\, r(t))dt + \sigma_1(t)dW_{1,t} \\ du &= -budt + \sigma_2(t)dW_{2,t}. \end{aligned}$$

Interest rates in this setup are again normally distributed, but the second stochastic factor u now allows for a rotation of the yield curve.

9.4 Market Models

Market models (also: *Libor* or *swap market models*) are conceptually different from short rate models. Many interest rate products, such as the loan example of Section 9.1, can be seen as derivatives on market interest rates (such as Libor or Euribor). One can also model these market rates directly, instead of taking the detour via short rates. The first step in this direction goes back to F. Black.[9]

Consider the price at time t of a caplet on the 6-month LIBOR at time T, with $t < T$. The discounted pay-off of the caplet is given by

$$D(T + 0.5) \cdot 0.5 \cdot (\text{Libor6M}(T) - K)^+,$$

where $D(T + 0.5)$ is the appropriate discounting factor. The *forward interest rate*

$$\text{Libor6M}_t(T) = \frac{1}{0.5} \log\left(\frac{Z(t; T)}{Z(t; T + 0.5)}\right),$$

where $Z(t; T)$ is the price at time t of a zero-coupon bond with principal 1 and maturity T, shall follow a geometric Brownian motion with drift μ and volatility σ, and remain constant after T, i.e. $\text{Libor6M}_{T+s}(T) = \text{Libor6M}(T)$ for $s \geq 0$.

[9]Fischer Black developed a formula for options on commodity futures based on the Black-Scholes model. This formula was taken up by practitioners for the pricing of floors and caps and proven in a mathematically rigorous way only some years later. Still, the model is typically referred to as the *Black76 model*.

We would like to determine the price C of a caplet at time $t = 0$. C can then be expressed as the expected discounted pay-off under the risk-neutral measure due the Fundamental Theorem of Asset Pricing , i.e.

$$C = 0.5\,\mathbb{E}^{\mathbb{Q}}[D(T+0.5)(\text{Libor6M}(T) - K)^{+}],$$

and $D(T+0.5)$ is again random. In contrast to the previous section, we now do not have a model for $D(t)$. We can work around this issue by evaluating $\text{Libor6M}_t(T)$ in terms of units of the zero-coupon bond with maturity $T+0.5$.[10] With this trick we can use today's bond price $Z(0, T+0.5)$ as discounting factor, which can then go in front of the expectation since it is an observable quantity today, and we arrive at

$$C = 0.5\,Z(0; T+0.5)\,\mathbb{E}^{\widetilde{\mathbb{Q}}}[(\text{Libor6M}(T) - K)^{+}],$$

where $\widetilde{Q}$ describes the dynamics of $\text{Libor6M}_t(T)$ in units of the bond price. It turns out that, under $\tilde{Q}$, $\text{Libor6M}(T) \sim e^{Y}$ with $Y \sim N(-\sigma^2 T/2, \sigma\sqrt{T})$,[11] so that it follows that $\text{Libor6M}(T)$ has the same distribution as the stock price in the Black-Scholes model (with $r = 0$). Thus, the price of the caplet can be written as

$$C = 0.5\,Z(0, T) \cdot (\text{Libor6M}_0(T)N(d_1) - KN(d_2)),$$

with

$$d_1 = \frac{\log(\text{Libor6M}_0(T)/K) + \sigma^2 T/2}{\sigma\sqrt{T}}, \qquad d_2 = d_1 - \sigma\sqrt{T},$$

where $\text{Libor6M}_0(T)$ is the observable forward interest rate.[12]
Swaptions can also be priced under the assumption of log-normally distributed swap rates.[13] The prices of caps, floors and swaptions are therefore quoted in terms of implied volatility, similar to stock options.

Starting in 1997 with work of Alan Brace, Dariusz Gatarek and Mark Musiela [11], the Black76 model has been studied in detail over recent years. During this time, the model has been put on a solid mathematical basis, so that some sources also refer to it as *BGM model* (or: *LIBOR market model* or *LMM model*).

---

[10] This technique is called *Change of Numeraire* and can be justified in a mathematically rigorous way.

[11] This corresponds exactly to the Black76 formula, as Black also set the expected return to 0.

[12] Note that to date no short-rate model has been found that is consistent with the assumption of log-normally distributed forward interest rates.

[13] Log-normal swap rates are consistent neither with log-normal forward rates, nor with known short-rate models.

It is a classical assumption that a yield curve with, for example, yearly nodes t_k, is available today (at $t = 0$). No-arbitrage arguments enable us to extract forward rates $F_k(0)$ as applied to the period $(t_k, t_{k+1}]$ for all k. With these initial values $F_k(0)$, the dynamics of the forward rates can be modeled to follow for $t_0 \leq t < t_1$ the dynamics,

$$dF_k(t) = \sigma_k(t)F_k(t)\sum_{j=1}^{k}\frac{\rho_{jk}\sigma_j(t)F_j(t)}{1+F_j(t)}dt + \sigma_k(t)F_k(t)dW_k(t).$$

In this market model, one will have as many Brownian motions as forward rates (which are pairwise correlated with correlation coefficients ρ_{jk}). One would naturally expect that forward interest rates far in the future (e.g. F_{25} and F_{26}), show a correlation close to 1. Once the model has been set up, the time intervals $[t_k, t_{k+1}]$ (and hence the yield curve nodes) no longer move, which means that we practically fix the calendar days. As time elapses, t will pass nodes t_j, after which F_j becomes obsolete and the sum in the above dynamics is shortened accordingly.

If the pricing of an interest rate instrument requires the distribution of the 5-year swap rate of 15th July 2021, one can run a Monte Carlo simulation (see Chapter 11) and compute the rate in each run based on the realizations of the five Libor12M forward rates (2021 $\rightarrow$ 2022), (2022$\rightarrow$ 2023),..., (2025$\rightarrow$ 2026), each for the 15th of July.

Libor market models are always high-dimensional. For example, if a model is based on the 3-month rates over a horizon of 50 years, one would have 200 Brownian motions. As a consequence, Monte Carlo simulation is often the only feasible option when pricing derivatives (see Chapter 11). Cancelation rights can be dealt with by the so-called *Longstaff-Schwartz techniques* (see references below). Libor market models require the volatilities σ_k and the correlation matrix (ρ_{jk}) as input parameters. Robust determination of the correlation matrix typically uses lower-dimensional approaches (e.g. 4 to 6 free parameters).

9.5 Key Takeaways, References and Exercises

Key Takeaways

After working through this chapter you should understand and be able to explain the following terms and concepts:

- Many fixed income instruments can be understood as derivatives on interest rates: bonds, caplets/floorlets, caps/floors, swaptions
- Short-rate models: $r = r_t$ can be modeled as an Itô process
- The Hull-White model: $dr_t = (a(t) - b(t)r_t)dt + \sigma(t)dW_t$ (mean-reverting process; constant parameters give the Vasiček model). Explicit pricing formulas exist for zero-coupon bonds and European calls/puts on them, r_t can become negative

- The Black-Karsinski model produces only positive short rates: $\log(r_t)$ follows the Hull-White model
- Market Models: directly model the dynamics of the forward rates (rather than going through the short rates). High-dimensional models, solution typically requires Monte Carlo methods

References

The Libor market model is mathematically rigorously introduced by Brace, Gaterek & Musiela in [11]. A comprehensive discussion of the topics that were presented in this chapter is given by Brigo & Mercurio [13], Filipović [33], Rebonato [64], and Shreve [72]. Longstaff-Schwartz techniques (for American put options) are explained in detail in the original article [53]. For multi-factor models, their implementation is non-trivial and requires the careful choice of the regression variables.

Exercises

1. Let $\sigma_i, i = 1, ..., 10$ be the Black76 volatilities for caps (on Libor6M), with identical strike prices and maturities of 1 to 10 years. The yield curve shall be observable for arbitrary maturities (nodes). Derive a recursive representation (as with bootstrapping) for the volatilities of the single caps, assuming that the two caplets of the same year have the same volatility.[14]
2. Swaption pricing formula in the Black76 model: let r be the discount rate (continuously compounded) for time T and let F be the forward swap rate of the T_1-year swap at maturity T of a payer swaption with strike price K. (In case of exercise, the swap would become effective at time T and run for T_1 years.) In analogy to the caplet formula, derive the price of a payer swaption in the Black76 model at time $t = 0$,

$$V(\text{payer swaption}) = \left[\frac{1 - \frac{1}{(1+(F/m))^{T_1 \cdot m}}}{F}\right] e^{-rT}\,(FN(d_1) - KN(d_2))$$

   with $d_1 = \frac{\log(F/K)+(\sigma^2/2)T}{\sigma\sqrt{T}}$, $d_2 = d_1 - \sigma\sqrt{T}$ and m payments per year from the fixed leg ($m = 1$ in the EUR case, and $m = 2$ for USD).
3. Compute the solution of the system (9.6).
4. Show that (9.11) holds. To do so, consider the pay-off of the put option on the bond at time T, and prove that this amount must be invested in a bond that earns interest at Libor, has maturity S and a coupon payment only at maturity.
5. How would you deal with a floating coupon in a one-factor short rate model if the floating rate is set *in arrears* (i.e. on the coupon payment day at the end of the coupon period)?
6. Show that the short rate r_t at time t in the Hull-White model (9.3) with constant $b(t) \equiv b$ and $\sigma(t) \equiv \sigma$ is normally distributed with mean $r_0\,e^{-bt} + \int_0^t e^{-b(t-s)}\,a(s)ds$ and variance $\frac{\sigma^2}{2b}(1 - e^{-2bt})$. (Hint: Apply the Itô-formula to the process $e^{bt}\,r_t$)

[14] In practice, caps are quoted with a single volatility (which is the one that is applied to all caplets). This way of pricing identical caplets is therefore inconsistent if they are included in caps of different volatilities. However, just quoting one volatility facilitates the comparison of various price offers in the market.

Exercises with Mathematica and UnRisk

7. Apply the UnRisk command `GetReferenceRates` to compute the 5-year swap rates from short rates between 4 and 10 % in a Vasiček model (as a particular case of the Hull-White model) with $\sigma = 0.01$, reversion speed $b = 0.1$ and $a = b \cdot 0.07$. How do the swap rates change, if the reversion speed is increased significantly? `GetReferenceRates` enables you to price interest rate instruments using simulation techniques (see Chapter 11) in the absence of cancelation rights.

Numerical Methods 10

Numerical techniques prove particularly useful when explicit solution formulas in a certain model cannot be derived even for simple derivatives (e.g. in the Black-Karasinski model) or when the to-be-priced financial instrument has a complex structure so that analytical methods fail (e.g. if multiple cancelation rights exist).

We will start this chapter by explaining algorithms that use binomial and trinomial trees (recall that we have already come across binomial structures when deriving the CRR model in Section 5.3). This will be followed by a discussion of numerical methods that can be applied to solving partial differential equations in financial applications (cf. Chapters 7, 8 and 9). Finally, we will outline an efficient numerical method to price European calls/puts when the characteristic function of the distribution of the log price of the underlying is known.

10.1 Binomial Trees

It is intuitive to use binomial trees for the pricing of option structures, if the Cox-Ross-Rubinstein model offers a suitable framework to model the price of the underlying. Consider now a CRR event tree as constructed in Section 5.3 (cf. Figure 5.1). The value of the option is known for all nodes (price realizations S_T) at maturity T. To price a European option, one can then recursively move backwards along the branches of the tree until the valuation date (or: initial date if $t_0 = 0$) is reached. For some time step size $\Delta t = T/n$, the corresponding Mathematica code could read as follows (stock price `S`, strike price `K`, interest rate `r`, volatility `sigma`, maturity `T`, number of time intervals `n`):

```
BinomialEuropeanCall[S_, K_, r_, sigma_, T_, n_] :=
  Module[{dt, a, up, down, P, Q, BinomTree, value,
            level},
    dt = T/n;
    a = Exp[r*dt];
    up = Exp[sigma*Sqrt[dt]]*a;
```

H. Albrecher et al., *Introduction to Quantitative Methods for Financial Markets*, Compact Textbooks in Mathematics, DOI 10.1007/978-3-0348-0519-3_10,

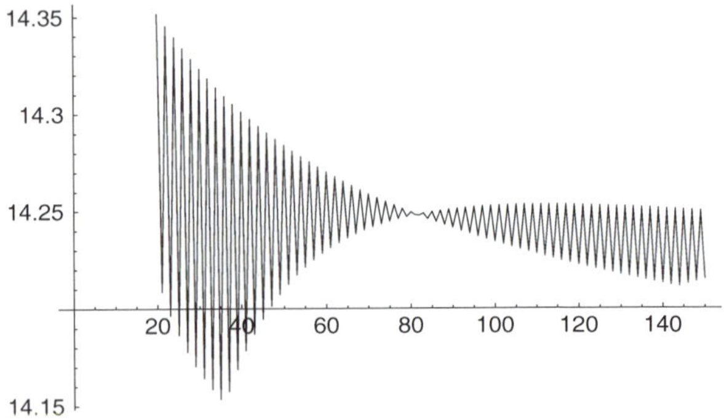

Fig. 10.1 Price of a vanilla European call option as a function of the number of time discretization points in a binomial tree

```
down = Exp[-sigma*Sqrt[dt]]*a;
P = (a - down)/(up - down);
Q = 1 - P;
P = P*Exp[-r*dt];
Q = Q*Exp[-r*dt];
BinomTree = Table[Max[S*down^node*up^(n - node)
                        - K, 0], {node, 0, n}];
Do[BinomTree =
     Table[{P, Q}.{BinomTree[[node]],
        BinomTree[[node + 1]]}, {node, 1, level}];
           {level, n, 1, -1}];
value = BinomTree[[1]];
Clear[BinomTree];
value]
```

The above binomial tree implementation in Mathematica does not give the computationally fastest way; however, it serves its purpose of illustrating the method. We now turn to the question of how the choice of a particular discretization step-size Δt affects the computed price of the European call option. Note that for $r = 0.05$, $\sigma = 0.3$, $T = 1$, $S = 100$ and $K = 100$ the Black-Scholes formula returns the fair price of the European call as 14.2313. Figure 10.1 shows the computed option prices for binomial trees using a discretization time grid of between $n = 20$ and $n = 150$ points. Looking at the plotted function, we immediately observe its oscillating nature depending on n being even or odd (see Exercise 2). There are other types of binomial tree setups that manage to greatly reduce these oscillations, which are due to the choice of the discretization grid (for example, by averaging the option prices for two neighboring numbers of time points n, e.g. $n = 40$ and $n = 41$).

Additionally, there are other issues around binomial trees that require attention. Consider a European up-and-out barrier option that is canceled as soon as the stock price exceeds a certain *barrier* by maturity T (see Chapter 2). Figure 10.2 depicts the option price as a function of the stock price for a discrete time grid of $n = 50$ steps over that week.

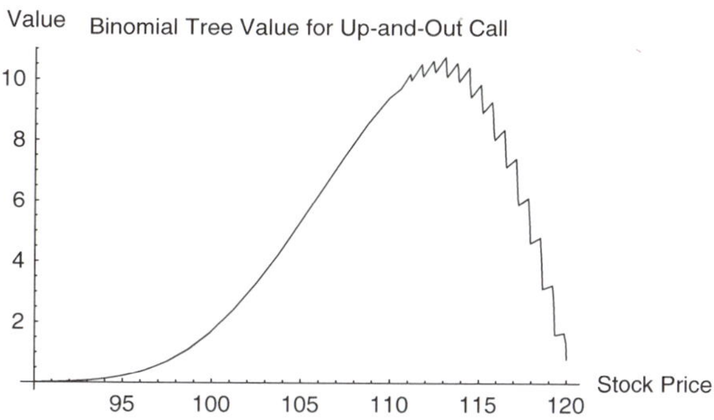

Fig. 10.2 Value of an up-and-out call as a function of the underlying stock price ($n = 50$ time discretization steps in the binomial tree, $K = 100$, barrier $B = 120$, $\sigma = 0.3$, $r = 0.05$ and $T = 7/365$)

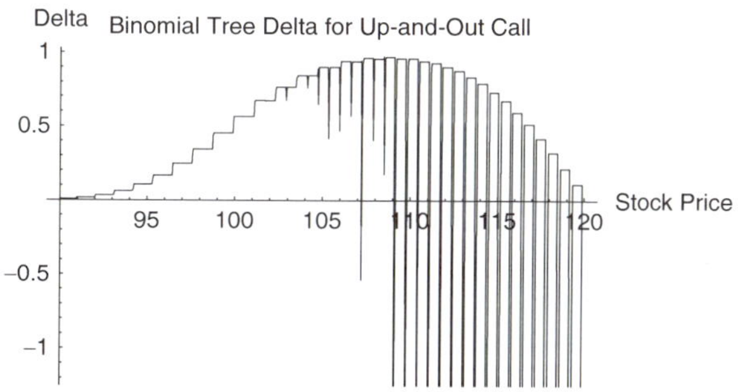

Fig. 10.3 Delta of the up-and-out call as function of the stock price when using binomial trees (50 steps)

Taking averages of 'odd' and 'even' trees does not provide a remedy for removing the spikes at the right end of the graph, although the sizes of the spikes can be reduced by narrowing the time grid, i.e. by increasing n. To list a third issue, the binomial tree method is also inadequate for determining the option delta $\frac{\partial V}{\partial S}$ for hedging purposes (see Figure 10.3): despite the actual Black-Scholes delta becoming negative when the stock price S comes sufficiently close to the barrier, the binomial tree still returns positive deltas for most S values (which is a result of the spiked shape as above). A naïve implementation of the delta-hedging strategy based on a binomial tree can therefore increase the portfolio risk rather than reduce it.

Algorithmic Benefits and Shortcomings of Binomial Trees

Binomial trees are straightforward and intuitive to implement and yield indicative and useful results in the absence of digital conditions[1]. Still, the shortcomings, such as the oscillating nature of the obtained option prices or the extreme errors in the option price sensitivities, mostly outweigh the advantages. Also, in the case of time-dependent interest rates or time-dependent volatility, it is no longer true that the

[1] Such a digital condition could be 'the option value remains 0, once the underlying stock price exceeds a certain barrier'.

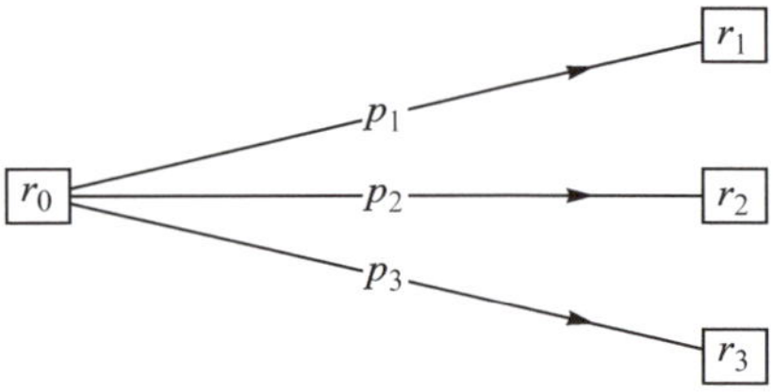

Fig. 10.4 Branch structure in a node of the trinomial tree

stock price at some later time only depends on the total number of up- and down-steps (i.e. that up-down and down-up steps would result in the same node). This can lead to an exponential complexity of the algorithm. Finally, the discretizations of time and stock price states cannot be conducted independently. A finer discretization of the stock prices also requires a finer time grid.

10.2 Trinomial Trees

For short-rate models, in particular for mean-reverting ones, trinomial trees can be applied to compute derivative prices. For trinomial trees, each node has exactly three successor nodes (instead of two as in the binomial model) with transition probabilities p_1, p_2, p_3 (see Figure 10.4).

For a particular state r_0, the three successor states r_1, r_2, r_3 should be a discrete approximation of the continuous state space of the distribution after one a time step Δt. Note that this is similar to the construction of binomial trees. Once the states r_1, r_2, r_3 are fixed, one has to assign the probability weights p_1, p_2, p_3 to the respective branches. It is natural to demand $p_1 + p_2 + p_3 = 1$, and with two more equations one can ensure that the mean and the variance of the trinomial model match the distribution of the underlying model (which starts in r_0). In contrast to stock price models, the status of the short rate drives the discounting factor, so that different discount rates will be used along different paths through the tree. The trapezoidal rule, for example, uses $\exp(-(r_0 + r_i)\Delta t/2)$ as discounting factor from r_i $(i = 1, 2, 3)$ to r_0.

In mean-reversion models, the weights p_1, p_2, p_3 will no longer be the same in all nodes, as we will observe a 'pull towards the center of the tree'. For sufficiently long maturities of the instruments, this can ultimately lead to the occurrence of negative weights p_i, in which case the situation $|p_1| + |p_2| + |p_3| > 1$ can cause instability and oscillations. For trinomial tree methods, one can define *down-branchings* and *up-branchings* to control the outer bounds of the tree. In down-branchings all branches at the upper end lead downwards, and in up-branchings all branches at the lower end lead upwards (see Figure 10.5).

A tree with up- and down-branchings will then structurally look like Figure 10.6. Table 10.1 will compare the results of a trinomial tree implementation for a two-factor Hull-White model (cf. (9.12)) with another numerical method.

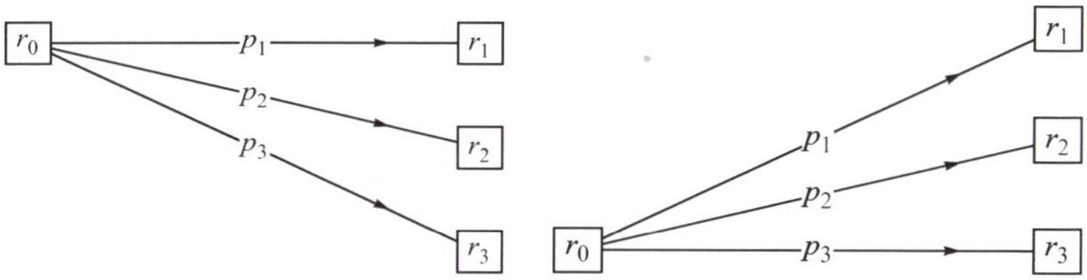

Fig. 10.5 Down-branching and up-branching

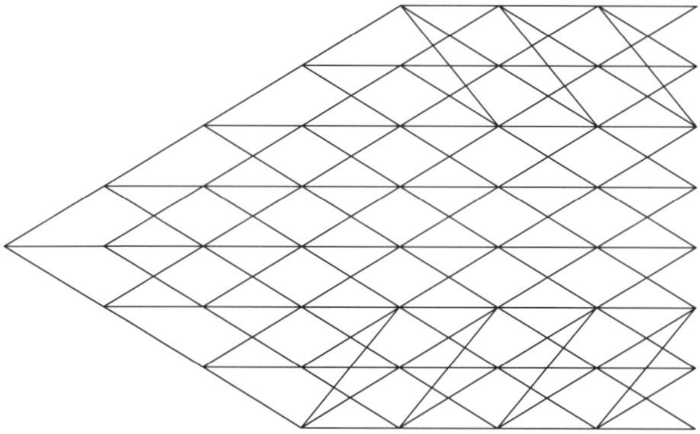

Fig. 10.6 Trinomial tree with regular, down- and up-branches

Table 10.1 Comparison of different pricing methods for a zero-coupon bond

maturity (years)	trinomial tree	analytic solution	FEM+streamline diffusion
1	0.950341	0.950353	0.950353
2	0.901665	0.901756	0.901756
4	0.808564	0.809135	0.809131
10	0.572741	0.576645	0.57662
20	0.315969	0.324704	0.324654
30	0.17375	0.18328	0.183224

10.3 Finite Differences and Finite Elements

Binomial and trinomial trees can also be interpreted as explicit numerical methods for the underlying partial differential equations. Taking a slightly different approach, one can try to employ numerical methods to solve the (parabolic) differential equation, for example by using finite differences. Recall the Black-Scholes differential equation (cf. Section 7.1),

$$\frac{\partial V}{\partial t} + \frac{\sigma^2 S^2}{2} \frac{\partial^2 V}{\partial S^2} + rS \frac{\partial V}{\partial S} - rV = 0. \tag{10.1}$$

Define a regular (equidistant) grid $S_i = i\Delta S$, $t_j = j\Delta T$, $V_{i,j} = V(S_i, t_j)$ on an appropriate region and replace the derivatives by difference quotients. The local discretization error (the approximation error in each time step) of the term

$$A_{i,j}^{\text{expl}} := \frac{V_{i,j} - V_{i,j-1}}{\Delta t} + \frac{1}{2}\sigma^2 S_i^2 \frac{V_{i+1,j} - 2V_{i,j} + V_{i-1,j}}{(\Delta S)^2} + rS_i \frac{V_{i+1,j} - V_{i-1,j}}{2\Delta S} - rV_{i,j}$$

is of order $O((\Delta S)^2, \Delta t)$. The *explicit difference method* now recursively solves the equations $A_{i,j} = 0$ with respect to $V_{i,j-1}$ as

$$V_{i,j-1} = V_{i,j} + \Delta t \left(\frac{1}{2}\sigma^2 S_i^2 \frac{V_{i+1,j} - 2V_{i,j} + V_{i-1,j}}{(\Delta S)^2} + rS_i \frac{V_{i+1,j} - V_{i-1,j}}{2\Delta S} - rV_{i,j} \right).$$

In order to ensure stability, explicit difference methods have to fulfill restrictive conditions regarding the time grid.[2] In our case stability is only ensured if

$$\Delta t \leq \frac{(\Delta S)^2}{\sigma^2 S^2}.$$

Hence, as one approaches equality in the above condition, a halving of the S step size requires to quarter the t step length, which then leads to 8-fold computational complexity.

To improve stability, one can turn to *implicit* or *Crank-Nicolson*-type methods. For implicit methods, a backward difference quotient is used to discretize the time steps,

$$A_{i,j}^{\text{impl}} := \frac{V_{i,j} - V_{i,j-1}}{\Delta t} + \frac{1}{2}\sigma^2 S_i^2 \frac{V_{i+1,j-1} - 2V_{i,j-1} + V_{i-1,j-1}}{(\Delta S)^2} + rS_i \frac{V_{i+1,j-1} - V_{i-1,j-1}}{2\Delta S} - rV_{i,j-1}.$$

This method is then *unconditionally stable*, i.e. there are no conditions requiring small time steps to ensure stability. However, a tri-diagonal equation system has to be solved now in each time step.

The *Crank-Nicolson method* (see the following discussion of the Finite Element method) combines the explicit and the implicit algorithm. Like the fully implicit method, it is stable at the expense of having to solve a linear equation system. With order of convergence $O((\Delta S)^2, (\Delta t)^2)$ it is significantly more efficient than the purely implicit method, which only has linear convergence in Δt.

[2] A detailed analysis of this aspect can be found in Zulehner [76].

Finite Elements

Finite differences are straightforward to implement in one dimension (here: for one underlying) and for equidistant grids. For higher-dimensional spaces and/or non-structured grids[3], *finite-element methods* (FEM) will often lead to better results. The fundamental idea here is the following. If (10.1) holds, it must also be true that

$$\int_0^\infty \left(\frac{\partial V}{\partial t} + \frac{\sigma^2 S^2}{2} \frac{\partial^2 V}{\partial S^2} + rS \frac{\partial V}{\partial S} - rV \right) \cdot W \cdot ds = 0$$

for arbitrary functions $W : [0, \infty) \to (-\infty, \infty)$ (the so-called *test functions*) for which the integration can be performed reasonably well.

Under the assumption that the following steps are permitted (i.e. under certain smoothness conditions), partial integration of the diffusion term $\frac{\sigma^2 S^2}{2} \frac{\partial^2 V}{\partial S^2}$ yields

$$\int_0^\infty \left(\frac{\sigma^2 S^2}{2} \frac{\partial^2 V}{\partial S^2} \right) \cdot W ds$$
$$= \frac{\sigma^2}{2} \left\{ \left[S^2 W \frac{\partial V}{\partial S} \right]_{S=0}^{S=\infty} - \int_0^\infty \left[2SW \frac{\partial V}{\partial S} + S^2 \frac{\partial W}{\partial S} \cdot \frac{\partial V}{\partial S} \right] dS \right\} .$$

If W has finite support, the first term above (i.e. $[S^2 W \frac{\partial V}{\partial S}]_{S=0}^{S=\infty}$) disappears for bounded $\frac{\partial V}{\partial S}(0)$. The remaining integral on the right must then be evaluated on the support of W only.[4] The Finite Elements method now applies a set of test functions W_i, which also serve as basis functions for V (here: only for the part that depends on S) and we decompose V with respect to this basis as

$$V(S, t_k) = \sum_i \alpha_{i,k} W_i(S).$$

One could, for instance, use piecewise linear basis functions W_i as depicted in Figure 10.7.

Again, starting with the terminating condition of the financial instrument, one runs backward in time. Assume in the following that $\alpha_{i,l}$ are known for $l > k$. In the backward time step from t_{k+1} to t_k, we compute

[3]Such non-structured grids could, for example, include triangular or tetrahedral grids, or grids that become finer in certain parts of the computational region.

[4]It will be practical if W only takes non-zero values on a small interval (see Figure 10.7). In this case the first term on the right-hand side of the above equation disappears, while the second (integral) term has a correspondingly small integration domain.

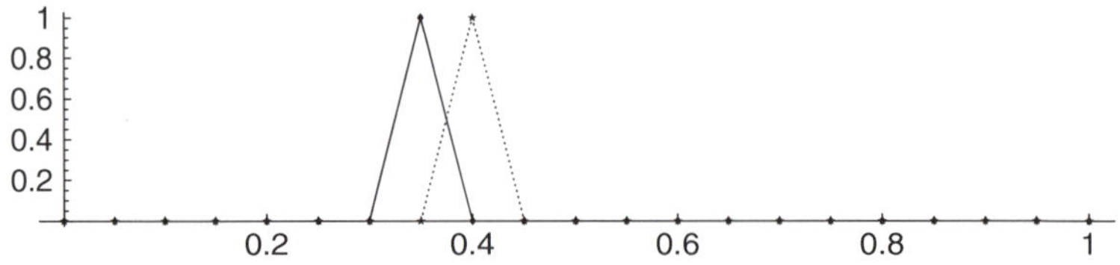

Fig. 10.7 Piecewise linear finite elements: two basis functions

$$\frac{\partial V}{\partial t} \approx \frac{1}{t_{k+1} - t_k} \sum_i (\alpha_{i,k+1} - \alpha_{i,k}) W_i, \qquad V \approx \xi \sum_i \alpha_{i,k} W_i + (1-\xi) \sum_i \alpha_{i,k+1} W_i,$$

$$\frac{\partial V}{\partial S} \approx \xi \sum_i \alpha_{i,k} \frac{\partial W_i}{\partial S} + (1-\xi) \sum_i \alpha_{i,k+1} \frac{\partial W_i}{\partial S}.$$

Hence, $\xi = 0$ corresponds to an explicit procedure, $\xi = 1$ leads to an implicit method, and $\xi = 1/2$ gives the Crank-Nicolson method.

The basis functions W_i are now chosen as test functions. Even when the basis functions are only piecewise linear, the first derivatives with respect to S exist almost everywhere. As we use partial integration, we do not require an explicit expression for the second partial derivatives. Note that in each time step the number of variables (the $\alpha_{i,k}$'s) equals the number of equations (one for each test function). In the explicit case, the system matrix is diagonal, in the fully implicit and Crank-Nicolson cases it is tri-diagonal. In the case where differential equations show significant *convection*, methods tailored for solving convection diffusion equations should be applied.[5]

Table 10.1 now compares the numerical results for the pricing of a zero-coupon bond in a two-factor Hull-White model with constant parameters ($\theta = 0.012, a = 0.2, b = 0.1, \sigma_1 = 0.01, \sigma_2 = 0.001, \rho = 0.3$) (see page 98) using (a) a doubly trinomial tree, (b) the analytic solution which is available in this example and (c) a finite element method with streamline diffusion to deal with the convection issue. The table shows that trinomial trees return low-quality results for longer maturities, while FEM produces results close to the explicit solutions in this example.

[5]In the Black-Scholes differential equation the term with the second derivative with respect to S is the *diffusion* term, which has smoothing properties. The term with the first derivative with respect to S is the so-called *convection term*. Heat transmission, for example, can occur by heat conduction (diffusion) or by the flow of fluids such as liquids or gases (convection). A central heating system in a house would be an example of convection. It is often challenging to treat convection numerically. The so-called upwind techniques or streamline diffusion techniques increase the stability of problems that show dominant convection. In the case of mean-reverting interest rate models, convection can be significant.

10.4 Pricing with the Characteristic Function

We discussed in Chapter 8 that the characteristic function of $\log S_t$ (under the risk-neutral measure $\mathbb{Q}$) can be obtained explicitly in many models. Recall that this was, for example, the case in the Heston model and the Merton model. The characteristic function fully determines the distribution of the random variable and can be applied to pricing European options. This shall now be discussed in more detail. Assume that $X_t = \log S_t$ has a (generally unknown) probability density function f_t under the risk neutral measure.[6] One can then write

$$\phi_t(z) = \mathbb{E}^{\mathbb{Q}}[e^{izX_t}] = \int_{-\infty}^{\infty} e^{izx} f_t(x) dx \tag{10.2}$$

for the characteristic function of the stock log price.

As European call and put options are actively traded, their prices can be used for calibration purposes. This underlines the importance of fast methods for the pricing of these vanilla options. We shall concentrate here on European calls, but the discussed method will equally work for put options (see Exercise 1). In principle one could apply inverse Fourier transformation to find the density function (which would then allow to determine the option price). This, however, mostly leads to integrals of singular integrands so that their evaluation is often numerically inefficient. The following trick allows for the efficient pricing of European calls. First, fix the maturity T of the option. The strike price, however, is *not* fixed up-front, and we interpret the call price C as a function of $k = \log K$. As $X_t = \log S_t$, one can write

$$C(k) = e^{-rT}\mathbb{E}^{\mathbb{Q}}[(e^{X_T} - e^k)^+] = e^{-rT} \int_k^{\infty} (e^x - e^k) f_T(x)\, dx.$$

Since the function C is well-defined on $\mathbb{R}$, one can compute its Fourier transform. We have that

$$\lim_{k \to -\infty} C(k) = S_0 > 0,$$

which is an immediate consequence of the no-arbitrage inequality $S_0 - e^k e^{-rT} \leq C(k) \leq S_0$. It follows that C is neither quadratically nor absolutely integrable, and so the Fourier transform cannot be computed in the classical sense. To overcome this issue, one introduces the new function $\gamma(k) := e^{\alpha k} C(k)$. If $\alpha > 0$ and $\mathbb{E}[e^{(1+\alpha)X_T}] < \infty$, we find that γ is bounded and integrable, so that the Fourier transform $\widehat{\gamma}$ is given as usual by

$$\widehat{\gamma}(z) = \int_{-\infty}^{\infty} e^{izk} \gamma(k)\, dk.$$

[6]This is the case for most stock price models discussed here, in particular for the Heston and the Merton model.

Conversely,

$$e^{\alpha k} C(k) = \gamma(k) = \frac{1}{2\pi} \int_{-\infty}^{\infty} e^{-ikz} \widehat{\gamma}(z)\, dz,$$

so that it remains to determine $\widehat{\gamma}(z)$,

$$\begin{aligned}
\widehat{\gamma}(z) &= \int_{-\infty}^{\infty} e^{izk} e^{\alpha k} C(k)\, dk = \int_{-\infty}^{\infty} e^{izk} e^{\alpha k} e^{-rT} \int_{k}^{\infty} (e^x - e^k) f_T(x)\, dx\, dk \\
&= e^{-rT} \int_{-\infty}^{\infty} f_T(x) \left(e^x \int_{-\infty}^{x} e^{(iz+\alpha)k}\, dk - \int_{-\infty}^{x} e^{(iz+\alpha+1)k}\, dk \right) dx \\
&= e^{-rT} \int_{-\infty}^{\infty} f_T(x) \left(\frac{e^x e^{(iz+\alpha)x}}{iz+\alpha} - \frac{e^{(iz+\alpha+1)x}}{iz+\alpha+1} \right) dx \\
&= e^{-rT} \int_{-\infty}^{\infty} f_T(x) e^{i(z-i(\alpha+1))x} \frac{1}{(iz+\alpha)(iz+\alpha+1)}\, dx \\
&= \frac{e^{-rT} \phi_T(z - i(1+\alpha))}{(iz+\alpha)(iz+\alpha+1)}.
\end{aligned}$$

Note that the order of integration was changed,[7] and the last equality follows from(10.2). $C(k)$ can hence be written as

$$\begin{aligned}
C(k) = e^{-\alpha k} \gamma(k) &= \frac{e^{-\alpha k}}{2\pi} \int_{-\infty}^{\infty} e^{-ikz} \frac{e^{-rT} \phi_T(z - i(1+\alpha))}{(iz+\alpha)(iz+\alpha+1)}\, dz \\
&= \frac{e^{-\alpha k}}{\pi} \int_{0}^{\infty} \mathrm{Re} \left(e^{-ikz} \frac{e^{-rT} \phi_T(z - i(1+\alpha))}{(iz+\alpha)(iz+\alpha+1)} \right) dz, \qquad (10.3)
\end{aligned}$$

where Re denotes the real part of a complex number and the last equality follows from the symmetry property of the real part of the integrand. Efficient numerical procedures exist to evaluate the final expression, including classical numerical integration procedures (e.g. Gauss quadrature) or the *Fast-Fourier transform* (FFT) (see Exercise 7). The latter allows to price an option in only one iteration if all model parameters and the maturity are given. Hence, FFT is a powerful tool to produce entire *option surfaces* (i.e. option prices for a large set of combinations of strikes and maturities) in a matter of seconds, which is very useful in the context of model calibration.

The above derivations hold for arbitrary $\alpha > 0$ as long as $\mathbb{E}[e^{(1+\alpha)X_t}] < \infty$, so that α is chosen with the aim of keeping the numerical integration efficient.

[7] This can be justified, as the integrand is absolutely integrable under the given assumption for α.

10.5 Numerical Algorithms in UnRisk

For the numerical treatment of derivatives in the Black-Scholes model, UnRisk applies numerical integration to

$$V(S_i, t^k) = \int_0^\infty G(S_i, \tilde{S}, t_k, t_{k+1}) V(\tilde{S}, t_{k+1}) d\tilde{S}$$

with

$$G(S, \widetilde{S}, t, T) = \frac{e^{-r(T-t)}}{\sigma \widetilde{S} \sqrt{2\pi(T-t)}} e^{\frac{-(\log(S/\tilde{S})+(r-\sigma^2/2)(T-t))^2}{2\sigma^2(T-t)}}$$

on some grid (S_i). The function $G(S, \widetilde{S}, t, T)$ is the probability density function of a log-normal distribution and the so-called *Green* function of the Black-Scholes differential equation. The representation is only valid for intervals (t_k, t_{k+1}) in which early exercise of the option is not permitted. For Bermudan options, it is checked at every possible exercise time and at every point on the grid whether the exercise value of the option lies above the retention value. UnRisk uses an analogous representation for one-factor short-rate models, and a numerical approximation of the Green function in the Black-Karasinki model. For the two-factor Hull-White model, finite elements and streamline diffusion are used.

Libor market models are solved by Monte Carlo methods that will be further discussed in Chapter 11. Note that it can also be comparably more efficient to apply Monte Carlo methods for specific cases of short-rate models, for example, where early exercise or redemption depends on some observable *trigger event* ('if a certain event occurs, a payment has to be made').

For advanced volatility models (e.g. Heston, Variance Gamma or Normal Inverse Gaussian models), UnRisk uses Fourier cosine methods (similar to FFT as described in the previous section) and/or Monte Carlo techniques.

10.6 Key Takeaways, References and Exercises

After working through this chapter you should understand and be able to explain the following terms and concepts:

- Binomial trees: structurally similar to CRR trees, option price as function of discretization step number n shows oscillations, option Delta inadequate for hedging
- Trinomial trees: used for short-rate models, same basic idea as binomial models, different discount factors along different branches
- Finite Difference methods: used to solve partial differential equations (e.g. the Black-Scholes PDE), we distinguish explicit (stability condition), implicit and Crank-Nicolson methods, straight forward to implement for one-dimensional problems

- Finite Elements methods: solve partial differential equation by the use of test functions, works also for higher-dimensional problems. Better performance than trinomial trees
- Characteristic functions: one obtains an integral expression (as function of the characteristic function) for $C(K)$, can be solved by numerical integration or FFT (can be very fast in producing option surfaces)

References

Trinomial trees have been suggested for the pricing in short-rate models by various authors, e.g. Hull & White [42, 43]. The presented method for the efficient pricing of European call options through characteristic functions was developed by Carr & Madan [15], so that (10.3) is also referred to as *Carr-Madan formula*. For further details on the FFT method, consult Lee [51] or Lord & Kahl [54]. Zulehner [76] is a good and comprehensive introduction to numerical aspects of partial differential equations, see also Larsson & Thomée [50]. Roos, Stynes & Tobiska [66] deal with numerical methods when convection is not negligible. Topper [73] applies the method of FFT to problems in financial mathematics. Binder & Schatz [6] use streamline diffusion and finite elements in particular for the two-factor Hull-White model. Comprehensive discussions of numerical solution strategies in financial mathematics are provided by Aichinger & Binder [1], Fusai & Roncoroni [36] and Seydel [70].

Exercises

1. In analogy to (10.3), produce a formula for the price of the corresponding put option. What are the restrictions on the choice of α?

Tasks with Mathematica

2. Check that the Mathematica code on page 103 is a realization of a multi-period Cox-Ross-Rubinstein tree (cf. Sections 5.3 and 7.2). Implement the code and reproduce the plot in Figure 10.1. Show that further increasing the number of time steps reduces the amplitudes of the oscillation around the correct value.
3. Further to Exercise 2, write down a code for pricing a European put option. How would the code for a European up-and-out call option differ?
4. Further to Exercise 2, what modifications would you have to make for pricing an American put option if you assume that the option can be exercised only at the times that are modeled by the binomial tree (which would practically correspond to a Bermudan put option)?
5. Write a program for the explicit difference method for an up-and-out call option with $S = 100$, $K = 100$, $B = 200$, $r = 0.05$, $\sigma = 0.3$ and $T = 1$. Discretize S equidistantly between 0 and 200 (with boundary condition $V = 0$). Plot the value for $t = 0$. What are the results for $\Delta S = 1$, $\Delta t = 1/100, 1/1000, 1/10000$?
6. Consider the Heston model of Chapter 8 with $\kappa = 0.6067$, $\theta = 0.0707$, $\lambda = 0.2928$, $\rho = -0.7571$ and $v_0 = 0.0654$. Further assume that $S_0 = 1$ and $r = 0.03$. Apply formula (10.3) with $\alpha = 0.75$ and the Mathematica command `NIntegrate`, to price a call with maturity $T = 1$ year and strike price $K = 1.1$.
7. The FFT algorithm allows us to determine the option prices for a set of strike prices in only one iteration. Assume that $S_0 = 1$. It can be shown that the application of *Simpson's rule* for numerical integration leads to the following approximation of the call price

$$C(k_l) \approx \frac{e^{-\alpha k_l}}{\pi} \sum_{j=1}^{n} e^{-2\pi(j-1)(l-1)/n}(-1)^{j-1}\psi_T(\eta(j-1))\frac{\eta}{3}(3+(-1)^j - 1_{\{j=1\}})$$

with $k_l = -\frac{\pi}{\eta} + l\frac{2\pi}{\eta n}, l = 1, \ldots, n.$

We can then compute the call prices for all n strikes in $n \cdot \log n$ steps. Set $n = 2^{12}$, $\eta = 0.125$ and use the Mathematics command `InverseFourier` to price the calls for all k_l using the parameters of the previous exercise. How will you choose the Mathematica parameter `FourierParameters`?

8. Assume that the stock price process under the physical probability measure follows the Heston model with parameters $\kappa = 5.13$, $\theta = 0.0436$, $\lambda = 0.52$ and $\rho = -0.754$. Assume further that $v_0 = \widetilde{\theta}$ in (8.14) (this assumption will ensure numerical stability), and that $S_0 = 1$ and $r = 0.03$. Plot the prices of European call options with maturity $T = 1$ year and strike price $K = 1.1$ as a function of the market price of risk ν. Note that the call prices will be a decreasing function of ν, since we start with the physical probability measure and choose the risk-neutral parameters in relation to ν.

Further numerical exercises will be provided in Chapter 13.

Simulation Methods 11

Many problems that arise in financial mathematics are structurally complex, so that one often cannot obtain explicit results (such as explicit pricing formulas for derivatives) or successfully apply numerical methods as outlined in Chapter 10. In such cases stochastic simulation can offer an efficient and powerful alternative for obtaining numerical estimates for specific quantities. The main part of this section will be dedicated to the Monte Carlo (MC) method and to ways of improving its computational efficiency, which can be very useful when implementing financial models in practice.

In general, (stochastic) *simulation* is about generating independent samples from a random variable $\mathbf{Z}$ (e.g. the random price $Z = S_T$ of a specific stock at a certain time T or a price path $\mathbf{Z} = (S_{t_1}, S_{t_2}, ..., S_T)$, cf. Figure 11.1).[1]

Once a set of N such sample points has been obtained, we can use this set to produce an *estimate* of some quantity that depends on the distribution of $\mathbf{Z}$, for example, the price of a knock-in call option when $\mathbf{Z}$ is the price path of the underlying. The quality of an estimate will be reflected by a *confidence interval* around this estimate.

11.1 The Monte Carlo Method

Recall that the value of a derivative can be written as the expectation of a function of some underlying random variable(s). Let $\mathbf{Z}$ be an s-dimensional random variable on $\mathbb{R}^s$ with cumulative distribution function $F_{\mathbf{Z}}(\mathbf{z})$ and $g : \mathbb{R}^s \to \mathbb{R}$ a real-valued function. We then define[2]

[1] In the sequel, possibly multi-variate random variables and vectors will denoted by **bold** letters or symbols.

[2] In the case of a European call option with strike K and maturity T we would set $Z = S_T$. The function $g(Z)$ would be the discounted pay-off $e^{-rT}(Z - K)^+$ of the option and F_Z the

H. Albrecher et al., *Introduction to Quantitative Methods for Financial Markets*, Compact Textbooks in Mathematics, DOI 10.1007/978-3-0348-0519-3_11,

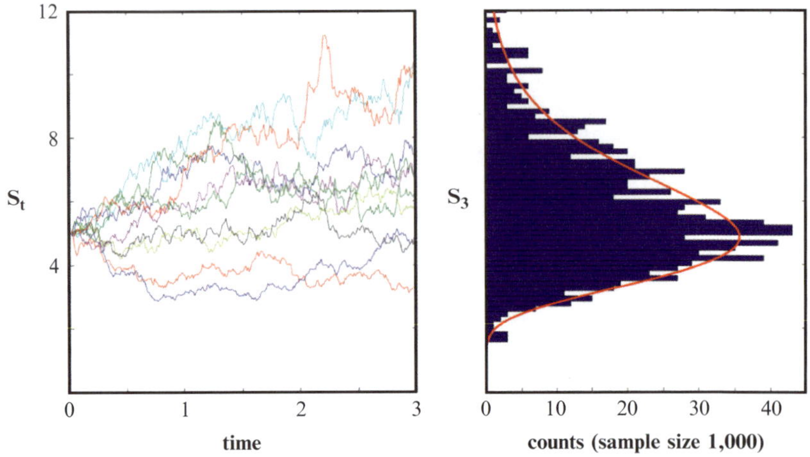

Fig. 11.1 10 sample paths of a geometric Brownian motion S_t on $[0, 3]$ (left) and histogram based on 1,000 sample points of the log-normal random variable S_3 (right), $S_0 = 5$, $\mu = 0.05$ and $\sigma = 0.2$

$$\alpha := \mathbb{E}[g(\mathbf{Z})] = \int_{\mathbb{R}^s} g(\mathbf{z})\, dF_{\mathbf{Z}}(\mathbf{z}). \tag{11.1}$$

If the above expectation cannot be computed explicitly, one has to employ numerical methods to obtain an estimate for it.[3] It is now intuitive to generate N independent sample points of $\mathbf{Z}$, i.e. we simulate $\mathbf{Z}$, and to approximate the true expectation by averaging over the set of generated sample points. This simple idea builds the foundation of the *Monte Carlo method*.[4]

(one-dimensional, i.e. $s = 1$) distribution function of the price S_T of the underlying asset under the risk-neutral measure.

[3]Let A be some event (e.g. that a random variable $\mathbf{Z}$ produces a negative realization, or that a sample path crosses a given barrier). We can then define the indicator function

$$\mathbf{1}_A(\mathbf{Z}) = \begin{cases} 1, & \text{if } \mathbf{Z} \in A, \\ 0, & \text{if } \mathbf{Z} \notin A. \end{cases}$$

As $\mathbb{P}(A) = \mathbb{E}[\mathbf{1}_A(\mathbf{Z})]$, a method to evaluate (11.1) will also allow to compute probabilities of events. In finance and insurance, such events of interest include the *default* of a bond (i.e. promised payments cannot be made in full), the *bankruptcy* of a company or the knock-in/knock-out event of a barrier option, to name a few.

[4]The Monte Carlo method was developed in the *Manhattan project* in the 1940s. The name is related to the randomness involved in the method, and finds its origin in the name of Monaco's administrative area Monte Carlo with its casino.

Example (Random number generation)

Mathematical software packages typically offer sophisticated methods for the generation of random numbers. In particular, being able to sample from a uniform distribution $U(0, 1)$ allows one to sample from a wide range of distributions.[5]

1. **Inversion method**. If $U \sim U(0, 1)$, then the random variable $F_Z^{-1}(U)$ has the same distribution as Z, where $F^{-1} := \inf\{z : F(z) \geq u\}$ is the inverse function of a cumulative distribution function. Hence, we can obtain a sample for any random variable Z from a sample of the uniform distribution.

   - **Exponential distribution**: for $Z \sim \text{Exp}(\nu)$ we have $F_Z(z) = 1 - \exp(-z\nu)$, so that $Z = -\frac{1}{\nu}\log(1 - U)$.
   - **Discrete distribution**: for $\mathbb{P}[Z = c_i] = p_i$ (discrete), $i = 1, ..., n$, set $q_0 := 0$ and $q_i := \sum_{j=1}^{i} p_j$, $i = 1, ..., n$. We obtain $Z = c_K$ for K satisfying $q_{K-1} < U \leq q_K$.

2. **Normal distribution N(0,1)**: an easy-to-implement method is the *Box-Muller method*. Sample from two independent random variables $U_1, U_2 \sim U(0, 1)$. Then both $Y_1 = \sqrt{-2\log(U_1)}\sin(2\pi U_2)$ and $Y_2 = \sqrt{-2\log(U_1)}\cos(2\pi U_2)$ are $N(0, 1)$ distributed.

3. ***d*-dimensional multivariate normal distribution $\mathbf{N}_d(\boldsymbol{\mu}, \Sigma)$**. Let $\mathbf{Z} \sim N_d(\boldsymbol{\mu}, \Sigma)$, with $\boldsymbol{\mu}$ being its mean vector and Σ its covariance matrix. For $\mathbf{X} \sim N_d(0, I_d)$ with identity matrix I_d (which requires sampling from d independent $N(0, 1)$ random variables), we have $\mathbf{Z} \sim A\mathbf{X} + \boldsymbol{\mu}$ with $\Sigma = AA'$, and $A = (a_{ij})$ is obtained from $\Sigma = (\sigma_{ij})$ by the *Cholesky decomposition*,

$$
\begin{aligned}
a_{ij} &= 0 \text{ for } j > i, a_{11} = \sqrt{\sigma_{11}}, \\
a_{ij} &= \left(\sigma_{ij} - \sum_{k=1}^{j-1} a_{ik}a_{jk}\right)/a_{jj} \text{ for } 1 \leq j < i \leq d, \\
a_{ii} &= \sqrt{\sigma_{ii} - \sum_{k=1}^{i-1} a_{ik}^2}.
\end{aligned}
$$

Note that the simulation of (geometric) Brownian motion will require the sampling from a normal distribution.

In general, $\mathbf{Z}$ can be multi-dimensional or even of high dimension. A suitable substitution allows one to translate the computation of (11.1) into the evaluation of

$$
\alpha = I(f) = \int_{[0,1]^s} f(\mathbf{x})\, d\mathbf{x}, \tag{11.2}
$$

where f is a real-valued function on the s-dimensional unit cube $[0, 1]^s$. One now chooses N random and independent integration points $\mathbf{x}_1, \ldots, \mathbf{x}_N$ in $[0, 1]^s$ (according to a uniform distribution on $[0, 1]^s$) and one approximates (11.2) by the arithmetic mean

[5] In practice, one will produce samples through a deterministic algorithm which imitates the uniform distribution well. This imitation is referred to as 'pseudo-random-number' algorithm and its quality can be assessed by statistical tests. The Mathematica command `RandomReal[1,n]` produces a set of n sample points from a $U(0, 1)$ distribution. For mathematical background on how to generate $U(0, 1)$ pseudo-random numbers see Korn et al. [48].

$$I_N(f) = \frac{1}{N}\sum_{n=1}^{N} f(\mathbf{x}_n). \tag{11.3}$$

Thanks to the *Strong Law of Large Numbers*, $I_N(f) \to I(f)$ for $N \to \infty$ with probability 1. The Central Limit Theorem then provides a tool to understand the approximation error

$$I_N(f) - I(f) = \frac{1}{N}\sum_{n=1}^{N} f(\mathbf{x}_n) - \mathbb{E}[f]$$

as an approximately normally distributed random variable with mean 0 and variance σ^2/N, where $\sigma^2 = \int_{[0,1]^s}(f(\mathbf{x}) - I(f))^2\,d\mathbf{x}$. The (asymptotic) 95 % confidence interval of $I_N(f)$ can therefore be estimated as

$$\left[I_N(f) - \frac{1.96\widehat{\sigma}}{\sqrt{N}},\, I_N(f) + \frac{1.96\widehat{\sigma}}{\sqrt{N}}\right],$$

with

$$\widehat{\sigma}^2 = \frac{1}{N}\sum_{n=1}^{N}(f(\mathbf{x}_n) - I_N(f))^2 = \frac{1}{N}\sum_{n=1}^{N}(f(\mathbf{x}_n))^2 - I_N(f)^2.$$

The Monte Carlo estimate $I_N(f)$ of $I(f)$ hence produces a probabilistic error bound of order $O(N^{-1/2})$. Note that this bound does not depend on the dimension s, as opposed to classical numerical integration methods.[6]

Variance-Reduction Methods

The fact that the approximation error of the Monte Carlo method is of (probabilistic) order $O(N^{-1/2})$ indicates that improving accuracy of an estimate by one decimal place requires an increase in the number of runs N by a factor 100.

To reduce the variance of the Monte Carlo estimator of α, one could, for example, aim to find another random variable $\mathbf{Z}'$ with $\mathbb{E}[g(\mathbf{Z}')] = \mathbb{E}[g(\mathbf{Z})] = \alpha$ and $\mathrm{Var}(g(\mathbf{Z}')) < \mathrm{Var}(g(\mathbf{Z}))$, so that an estimate of $\mathbb{E}[g(\mathbf{Z})]$ can be obtained by simulating $\mathbf{Z}'$. Finding $\mathbf{Z}'$ is a classical and often challenging problem in simulation. It will only be worthwhile to search for and implement a particular variance reduction algorithm if $\mathrm{Var}(g(\mathbf{Z}'))$ is significantly smaller than $\mathrm{Var}(g(\mathbf{Z}))$. If, for example, $\mathrm{Var}(g(\mathbf{Z}')) = \mathrm{Var}(g(\mathbf{Z}))/2$, N runs under this reduced-variance

[6]This is the reason why Monte Carlo methods are typically preferred over numerical integration methods in dimensions $s \geq 5$.

algorithm will achieve the same accuracy as simply running the 'crude' Monte Carlo algorithm $2N$ times. The improvement in efficiency hence needs to be compared to the extra time and effort of identifying $\mathbf{Z}'$ and implementing the variance-reduction algorithm.

In the following we will discuss three examples of widely-used variance-reduction strategies.

Conditional Monte Carlo: let $\mathbf{Y}$ be another random variable that is produced at the same time as $\mathbf{Z}$. Define $\mathbf{Z}' = g_1(\mathbf{Y}) = \mathbb{E}[g(\mathbf{Z})|\mathbf{Y}]$, so that $\mathbb{E}[g_1(\mathbf{Y})] = \mathbb{E}[g(\mathbf{Z})] = \alpha$, and $g_1(\mathbf{Y})$ can now also be used to produce a Monte Carlo estimate of α. Since

$$\mathrm{Var}(g(\mathbf{Z})) = \mathrm{Var}(\mathbb{E}[g(\mathbf{Z})|\mathbf{Y}]) + \mathbb{E}[\mathrm{Var}(g(\mathbf{Z})|\mathbf{Y})],$$

it follows that

$$\mathrm{Var}[g_1(\mathbf{Y})] \leq \mathrm{Var}[g(\mathbf{Z})],$$

so that Conditional Monte Carlo always leads to a reduction in variance.

Importance Sampling: here the idea is to use a different probability measure in the simulation, so that 'important' sample points (or: paths) are more likely to occur. For simplicity, assume that $\mathbf{Z}$ has a probability density function $f : \mathbb{R}^s \to \mathbb{R}_+$ (the more general case can be treated analogously).

It follows that

$$\alpha = \mathbb{E}[g(\mathbf{Z})] = \int_{\mathbb{R}^s} g(\mathbf{z})\, f(\mathbf{z})\, d\mathbf{z}.$$

Let $f_I(\mathbf{z}) : \mathbb{R}^s \to \mathbb{R}_+$ be a second probability density function with the property that $f(\mathbf{z}) > 0 \Rightarrow f_I(\mathbf{z}) > 0$ for all $\mathbf{z} \in \mathbb{R}^s \setminus \{\mathbf{z} : g(\mathbf{z}) = 0\}$. The integral

$$\alpha = \int_{\mathbb{R}^s} g(\mathbf{z})\, \frac{f(\mathbf{z})}{f_I(\mathbf{z})}\, f_I(\mathbf{z})\, d\mathbf{z}$$

can then be understood as an expectation with respect to the density f_I, i.e.

$$\alpha = \mathbb{E}_I\left(g(\mathbf{Z})\frac{f(\mathbf{Z})}{f_I(\mathbf{Z})}\right),$$

where the index I of the expectation indicates that $\mathbf{Z}$ is now distributed according to the density f_I.[7] One can then use Monte Carlo simulation to generate N sample points $\mathbf{z}_1, \ldots, \mathbf{z}_N$ from f_I and use the unbiased estimator

[7] $f(\mathbf{Z})/f_I(\mathbf{Z})$ is the so-called *Radon-Nikodým derivative* or *likelihood ratio*.

$$\hat{\alpha}_I = \frac{1}{N}\sum_{i=1}^{N} g(\mathbf{z}_i)\frac{f(\mathbf{z}_i)}{f_I(\mathbf{z}_i)}. \qquad (11.4)$$

This method provides variance reduction if f_I is chosen such that

$$\mathbb{E}_I\left(\left(g(\mathbf{Z})\frac{f(\mathbf{Z})}{f_I(\mathbf{Z})}\right)^2\right) = \mathbb{E}\left(g^2(\mathbf{Z})\frac{f(\mathbf{Z})}{f_I(\mathbf{Z})}\right) < \mathbb{E}(g^2(\mathbf{Z})).$$

Identity (11.4) implies that the variance of the estimator will be particularly low if $f_I(\mathbf{z})$ can be chosen close to proportional to the product $g(\mathbf{z})f(\mathbf{z})$. If g describes, for instance, the pay-off of an option and f is the risk-neutral density, then it would be desirable for f_I to have more probability mass (and, therefore, produce more sample realizations) in the regions where this product is large. In the case of knock-in barrier options, f_I will be chosen such that the barrier is reached more often than under the original probability density f (see Exercise 11).

Typical ways of choosing the probability density function $f_I(\mathbf{z})$ include *shifting* ($f_I(\mathbf{z}) = f(\mathbf{z}+\boldsymbol{\theta})$, $\boldsymbol{\theta} \in \mathbb{R}^s$), *scaling* ($f_I(\mathbf{z}) = \frac{1}{\theta}f(\mathbf{z}/\theta)$, $\theta \in \mathbb{R}_+$) and *exponential twisting* ($f_I(\mathbf{z}) = \frac{\exp(\boldsymbol{\theta}^{\mathrm{T}}\mathbf{z})}{\mathbb{E}[\exp(\boldsymbol{\theta}^{\mathrm{T}}\mathbf{z})]}f(\mathbf{z})$, $\boldsymbol{\theta} \in \mathbb{R}^s$), for suitable $\boldsymbol{\theta}$.

Control Variates: consider a random variable Z with dimension $s = 1$ and, without loss of generality, define $g(z) \equiv z$. Furthermore, let Y be a random variable that has known mean $\mathbb{E}[Y]$ and is correlated to Z. (Z, Y) can then be simulated jointly and it clearly holds that

$$\mathbb{E}[Z] = \mathbb{E}[Z - a \cdot Y] + a\mathbb{E}[Y],$$

for constant $a \neq 0$. One now has to estimate $\mathbb{E}[Z - a \cdot Y]$ instead of $\mathbb{E}[Z]$. Produce N independent samples $(z_1, y_1), \ldots, (z_N, y_N)$ of the vector (Z, Y). Calculate

$$z_j(a) = z_j - a(y_j - \mathbb{E}[Y]), \quad j = 1, \ldots, N$$

and the new estimator

$$\hat{\alpha}_Z(a) := \frac{1}{N}\sum_{j=1}^{N} z_j(a) = \hat{\alpha}_Z - a(\hat{\alpha}_Y - \mathbb{E}[Y])$$

is computed based on the original Monte Carlo estimator $\widehat{\alpha}_Z = \frac{1}{N}\sum_{j=1}^{N} z_j$, 'corrected' for the observed estimation error $\widehat{\alpha}_Y - \mathbb{E}[Y] = \frac{1}{N}\sum_{j=1}^{N} y_j - \mathbb{E}[Y]$. It is obvious that $\mathbb{E}[\widehat{\alpha}_Z(a)] = \mathbb{E}[Z]$. Let σ_Y^2 and σ_Z^2 be the variances of Y and Z, respectively, and denote the correlation coefficient of Y and Z by $\rho_{Y,Z}$. We then find

$$\mathrm{Var}(Z_j(a)) = \sigma_Z^2 - 2a\sigma_Y\sigma_Z\rho_{Y,Z} + a^2\sigma_Y^2 = N\ \mathrm{Var}(\widehat{\alpha}_Z(a)). \qquad (11.5)$$

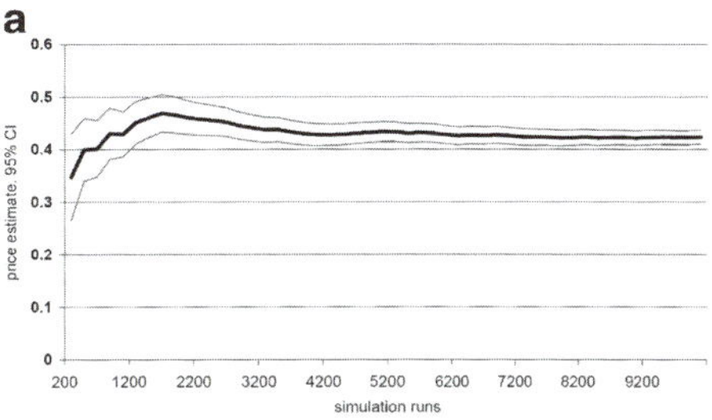

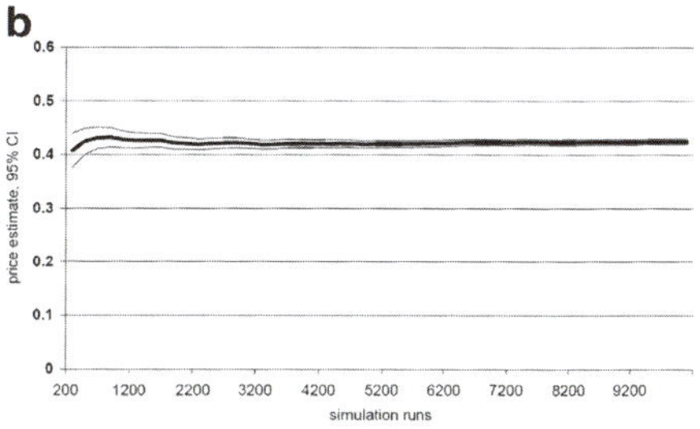

Fig. 11.2 Simulated call option prices based on (a) crude Monte Carlo (left) and (b) using another call option with known price as control variate (right), 10,000 simulation runs, 95% confidence intervals

Since $\mathrm{Var}(\hat{\alpha}_Z) = \mathrm{Var}(\widehat{\alpha}_Z(0)) = \sigma_Z^2/N$, the new estimator $\widehat{\alpha}_Z(a)$ has smaller variance than $\widehat{\alpha}_Z$ if $a^2\sigma_Y < 2a\sigma_Z\,\rho_{Y,Z}$. In particular, (11.5) allows one to determine the value of a that minimizes the variance of the new estimator as

$$a^* = \rho_{Y,Z}\frac{\sigma_Z}{\sigma_Y} = \frac{\mathrm{Cov}(Y, Z)}{\mathrm{Var}(Z)}.$$

The improvement in variance is hence given by

$$\frac{\mathrm{Var}(\widehat{\alpha}_Z(a^*))}{\mathrm{Var}(\widehat{\alpha}_Z)} = 1 - \rho_{Y,Z}^2,$$

so that stronger correlation between Z and Y leads to lower variance. In practice, $\mathrm{Var}(Z)$ and $\rho_{Y,Z}$ will be unknown (as the aim is to estimate $\mathbb{E}[Z]$); however, the procedure typically works well if a pre-simulation is run to estimate a^* as

$$\widehat{a}^* = \frac{\sum_{j=1}^n (y_j - \widehat{\alpha}_Y)(z_j - \widehat{\alpha}_Z)}{\sum_{j=1}^n (y_j - \widehat{\alpha}_Y)^2}.$$

The control-variate method often leads to significant variance reduction and is used in various areas of quantitative finance, for example, for the pricing of options whose pay-off is similar to another option for which an explicit pricing formula is known, as illustrated in the following example.

Example (Control variates)

We will now compare the performance of employing the control-variate technique in option pricing vs. crude Monte Carlo simulation. Using simulation (based on a Black-Scholes model), we aim at determining the price $C_0^{1.3}$ of a call option with strike $K = 1.3$ and maturity $T = 3$ on an underlying stock with initial stock price $S_0 = 1.4$ and volatility $\sigma = 0.3$ (which by the Black-Scholes formula is 0.413). Assume the risk-free interest rate is $r = 0.05$. The price $C_0^{0.8}$ of an otherwise identical option, but with strike $K_2 = 0.8$, is known to be 0.731. After determining the optimal multiplier $a^* = 0.577$, Figure 11.2 shows that the control-variate method (using the call with strike 0.8 as control variate) significantly outperforms crude Monte Carlo simulation. After 4,000 runs the (asymptotic) 95% confidence bounds almost run with the estimate, while even after 10,000 runs the estimate and the confidence bounds in the crude Monte Carlo method are not yet narrow.

11.2 Quasi-Monte Carlo (QMC) Methods

As previously mentioned, the classical Monte Carlo method produces an approximation error of the order $O(N^{-1/2})$ which is probabilistic. This motivates the use of deterministic point sequences, which preserve the advantages of classical Monte Carlo, but also provide deterministic, and asymptotically better, error bounds.

Instead of using random points in $[0, 1]^s$ for the computation of (11.2), one hence uses deterministic point sequences which are known to imitate the properties of the uniform distribution well. When looking for a measure of how well a given sequence $\{\mathbf{x}_n\}_{n=1}^N$ mimics a uniform distribution on I^s, one can define the so-called *star discrepancy*

$$D_N^*(\mathbf{x}_1, \ldots, \mathbf{x}_N) = \sup_{\boldsymbol{\beta} \in [0,1]^s} \left| \frac{1}{N} \sum_{n=1}^{N} 1_{[0,\boldsymbol{\beta})}(\mathbf{x}_n) - \beta_1 \cdot \beta_2 \cdots \beta_s \right|,$$

with $[0, \boldsymbol{\beta}) = [0, \beta_1) \times \ldots \times [0, \beta_s)$ and 1_A being the indicator function of the set A. A sequence $\omega = \{\mathbf{x}_n\}_{n=1}^{\infty}$ is called *uniformly distributed*, if

$$\lim_{N \to \infty} D_N^*(\omega) = 0.$$

The theory of uniform distribution proves that $I_N \to I$ as $N \to \infty$ for uniformly distributed sequences. An upper bound of the approximation error is given by the so-called *Koksma-Hlawka inequality*,

$$\left| I - \frac{1}{N} \sum_{n=1}^{N} f(\mathbf{x}_n) \right| \leq V(f) D_N^*(\omega), \tag{11.6}$$

where $V(f)$ is the variation[8] of the function f. Smaller discrepancy values hence lead to lower approximation errors. Note that the error bound can be decomposed into one component which solely depends on the properties of the integrand f, while the other term $D_N^*(\omega)$ only depends on the used sequence. The star discrepancy of the best known sequences has asymptotic order $O((\log N)^s/N)$, and such sequences are called *low-discrepancy sequences*. Due to (11.6), the application of such sequences gives an approximation error of order $O((\log N)^s/N)$. In contrast

[8]Concretely, it is the total variation in the sense of Hardy and Krause.

to the Monte Carlo method, this error bound now depends on s while being deterministic. Also note that this bound describes a 'worst case', and typically integration errors in the application will be significantly lower than this bound. A couple of low-discrepancy sequences are discussed in the following.

Van der Corput and Halton Sequences

Choose some integer $b \geq 2$, and write an integer $n \geq 0$ as its unique base-b representation $n = \sum_{j=0}^{\infty} a_j(n)b^j$ (with $a_j(n) \in \{0, \ldots, b-1\}$). The n-th term x_n of the *Van der Corput sequence with base b* is then given by reflecting this representation at the decimal point, i.e.

$$x_n := \phi_b(n) = \sum_{j=0}^{\infty} a_j(n)b^{-j-1}.$$

This construction can easily be extended to $s > 1$ dimensions, as one can choose s relatively prime integers $b_1, \ldots, b_s \geq 2$. The *Halton sequence with bases* $b_1, \ldots, b_s$ is given by

$$\mathbf{x}_n = (\phi_{b_1}(n), \ldots, \phi_{b_s}(n)) \in I^s \qquad \text{for all } n \geq 0.$$

Nets have even lower discrepancy. A (t, m, s)*-net with base b* is defined as a set P of $N = b^m$ points in $[0, 1]^s$, such that every elementary interval

$$E = \prod_{i=1}^{s} [a_i b^{-d_i}, (a_i + 1)b^{-d_i}), \qquad a_i, d_i \in \mathbb{Z},\ d_i \geq 0,\ 0 \leq a_i < b^{d_i},\ 1 \leq i \leq s,$$

with $\text{vol}(E) = b^{t-m}$ contains exactly b^t points of P. Correspondingly, a sequence $\mathbf{x}_0, \mathbf{x}_1, \ldots$ of points in I^s is a (t, s)*-sequence with base b*, if the set of points $P_{k,m} = \{\mathbf{x}_n : kb^m \leq n < (k+1)b^m\}$ is a (t, m, s)-net for all $k \geq 0$ and $m > t$. For example, the van der Corput sequence with base b is a $(0, 1)$-sequence with base b. (t, s)-sequences with base 2 are called *Sobol sequences*. Their construction is more cumbersome than that of Halton sequences, but Sobol sequences are often found well-suited for integrands arising in problems in financial mathematics (see Exercise 17).

Figure 11.3 illustrates the distribution of the first 1,000 points of a two-dimensional sequence for (a) pseudo-random numbers, (b) a Halton sequence with bases 2 and 3 and (c) a Sobol sequence. Note that the QMC sequences fill the unit square more evenly than the pseudo-random sequence.

The distributional properties of Quasi-Monte Carlo sequences in few dimensions are generally found to be very good, and they mostly perform significantly better than classical Monte Carlo sequences. However, as the number s of dimensions increases, the error bound (11.6), as well as the actual approximation error, can become very large. In practice, one can try to overcome this issue for high-dimensional problems by the use of QMC points for the first few (e.g. 20)

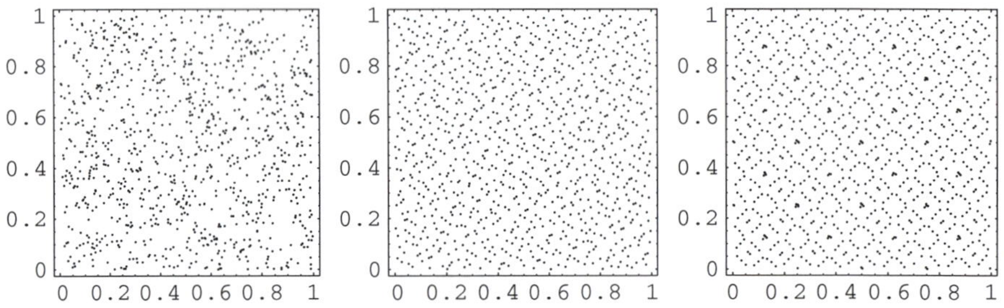

Fig. 11.3 Sequence of pseudo-random numbers (left), Halton sequence with bases $b_1 = 2$, $b_2 = 3$ (middle) and a Sobol sequence (right) in $[0, 1]^2$

dimensions, while sampling the points for the remaining dimension by classical Monte Carlo (such methods are called *hybrid methods*). Additionally, in integration problems some dimensions will often play a more 'important' role than others, and using these important dimensions as QMC-dimensions can significantly improve simulation efficiency. This is illustrated in the following example.

Example (Brownian motion)

A sample path of a Brownian motion W_t over the interval $[0, T]$ can be simulated as follows. Define a time step size $\Delta t = t_{n+1} - t_n = T/M$, and start every sample path in $W_0 = 0$. At each time t_{n+1} ($n = 0, \ldots, M - 1$), add the increment $W_{t_{n+1}} - W_{t_n}$ (by sampling from a normal distribution with mean 0 and variance T/M) to the already determined value W_{t_n}.[9] Now let $M = 2^m$ be some power of 2 and use a QMC to produce the sample points. As every dimension is 'equally important' in this algorithm, it is not exploited that the first dimensions of QMC sequences are 'better' distributed than the others. The *Brownian bridge* construction offers an alternative way of generating sample paths of Brownian motion. Sample W_T from $N(0, T)$. Generate a sample point of $W_{T/2}$ conditioned on the values $W_0 = 0$ and W_T, with $W_{T/2} \sim N(W_T/2, T/4)$ (see Exercise 5). In a next step, the realizations of $W_{T/4}$ and $W_{3T/4}$ are determined, which are again normally distributed with parameters dependent on the previously produced values. Continuing this procedure, the sample path is refined further and further, and the first few dimensions of the point sequence are used for the 'more important' points, i.e. the ones that will have the strongest impact on the outcome of the sample path.

Although again N points of an M-dimensional point sequence are required for simulation through Brownian bridges, the re-structuring of the simulation algorithm can significantly improve the efficiency of the Quasi-Monte Carlo method (see Exercise 18).

[9] For $T = 1$ and a time step $\Delta t = 0.01$, one will have to sample $M = 100$ normally distributed random variables to generate one sample path. Thus, the simulation of the price of an option with $N = 1{,}000$ sample paths will require 1,000 points of a 100-dimensional point sequence.

11.3 Simulation of Stochastic Differential Equations

Recall that the stochastic differential equation (6.8) for the geometric Brownian motion in Chapter 6 led to an explicit analytical solution. However, for many other processes used in financial mathematics, such explicit solutions are not available. In these cases one can turn to a numerical approximation scheme that allows to simulate the paths of the solution. Consider a general Itô process of the type

$$dX_t = \mu(X_t)\,dt + \sigma(X_t)\,dW_t. \tag{11.7}$$

Discretizing the time axis and defining a step size of Δt leads to the following approximation of (11.7):

$$X_{t+\Delta t} \approx X_t + \mu(X_t)\,\Delta t + \sigma(X_t)\,\Delta W_t. \tag{11.8}$$

Note that $\mu(X_\tau)$ and $\sigma(X_\tau)$, $t \le \tau \le t + \Delta t$, have been replaced by the respective values $\mu(X_t)$ and $\sigma(X_t)$ at the left end of the time interval. From the definition of Brownian motion it follows that $\Delta W_t \sim N(0, \Delta t)$. This approximation scheme (11.8) is called *Euler scheme*[10] and is a simple way of approximating solution paths of (11.7). After fixing the initial value X_0 and the time step size Δt, it is straightforward to generate a normally distributed random variable ΔW_t in each step, and to derive a path via (11.8). Under mild conditions on the functions μ and σ, the method can be shown to converge to the exact solution for $\Delta t \to 0$.

The above method can be improved as follows. From Itô's Lemma (6.7) it follows that

$$d\sigma(X_t) = \left(\sigma'(X_t)\,\mu(X_t) + \frac{1}{2}\sigma''(X_t)\,\sigma^2(X_t)\right)\,dt + \sigma'(X_t)\,\sigma(X_t)\,dW_t.$$

This equation for $s > t$ can again be discretized as above:

$$\begin{aligned}\sigma(X_s) - \sigma(X_t) = &\left(\sigma'(X_t)\,\mu(X_t) + \frac{1}{2}\sigma''(X_t)\,\sigma^2(X_t)\right)\,(s-t)\\ &+ \sigma'(X_t)\,\sigma(X_t)\,(W_s - W_t),\end{aligned}$$

where the coefficients are once again replaced by the values for the left end point t of the interval $[t, s]$. As $W_s - W_t$ is of magnitude $\sqrt{s-t}$ (which dominates $s-t$ for small intervals, see Chapter 6), we approximate

$$\sigma(X_s) \approx \sigma(X_t) + \sigma'(X_t)\,\sigma(X_t)\,(W_s - W_t).$$

For $X_{t+\Delta t}$ it follows that

[10] As it corresponds to the Euler approximation for the numerical solution of ordinary differential equations.

$$X_{t+\Delta t} \approx X_t + \int_t^{t+\Delta t} \mu(X_s)\, ds + \int_t^{t+\Delta t} \sigma(X_t) dW_s$$
$$+ \sigma'(X_t)\,\sigma(X_t) \int_t^{t+\Delta t} (W_s - W_t) dW_s.$$

Using (6.11), we can write

$$\int_t^{t+\Delta t} (W_s - W_t) dW_s = \frac{1}{2}(W_{t+\Delta t} - W_t)^2 - \frac{1}{2}\,\Delta t = \frac{1}{2}(\Delta W_t)^2 - \frac{1}{2}\,\Delta t,$$

so that we arrive at the *Milstein scheme*, which converges faster than the classical Euler scheme:

$$X_{t+\Delta t} \approx X_t + \mu(X_t)\,\Delta t + \sigma(X_t)\,\Delta W_t + \frac{\sigma'(X_t)\,\sigma(X_t)}{2}((\Delta W_t)^2 - t). \quad (11.9)$$

11.4 Key Takeaways, References and Exercises

Key Takeaways

After working through this chapter you should understand and be able to explain the following terms and concepts:

- Sampling from exponential, discrete and (multivariate) normal distributions
- The Monte Carlo method: expectation as integral, point estimate $\widehat{\alpha}$, (asymptotic) confidence interval
- Variance-reduction methods: Conditional Monte Carlo, Importance Sampling, Control Variates
- Quasi-Monte Carlo methods: deterministic sequences to mimic randomness, discrepancy, van der Corput/Halton sequences and nets
- Brownian bridge construction of a Brownian motion
- Simulating SDEs: Euler scheme, Milstein scheme

References

A comprehensive overview of Monte Carlo methods applied to finance is given by the books of Glasserman [38] and Korn et al. [48]. Asmussen & Glynn [3] offer an abundant selection of techniques and ideas around stochastic simulation. Niederreiter [59] provides an introductory text to Quasi-Monte Carlo methods; the more advanced reader might wish to consult the book of Drmota & Tichy [22]. For more tools to simulate from multivariate distributions (e.g. by the use of copulas), see McNeil et al. [56].

Exercises

1. Show for $U_1, U_2 \sim U(0,1)$ that $Y_1 = \sqrt{-2\log(U_1)}\sin(2\pi U_2)$ and $Y_2 = \sqrt{-2\log(U_1)}\cos(2\pi U_2)$ are indeed $N(0,1)$ distributed. (Hint: use polar coordinates.)
2. Determine the factor that describes the reduction of variance of the Monte Carlo estimator for the control-variate technique, where Y is a random variable of known mean that is strongly correlated to Z; the correlation coefficient shall be 0.99, 0.98, 0.95 or 0.85.
3. Another variance-reduction method is the *antithetic method*. Instead of (11.3), one uses the estimator

$$I_N^A(f) = \frac{1}{2N}\sum_{n=1}^{N}\Big(f(\mathbf{x}_n) + f(\mathbf{1}-\mathbf{x}_n)\Big)$$

with $\mathbf{1} = (1,\dots,1)$. Prove that

$$\mathrm{Var}(I_N^A(f)) = \frac{1}{2N}\Big(\mathrm{Var}(f(\mathbf{U}_s)) + \mathrm{Cov}(f(\mathbf{U}_s), f(\mathbf{1}-\mathbf{U}_s))\Big),$$

where $\mathbf{U}_s$ is an s-dimensional random variable with independent and [0,1]-uniformly distributed components, such that the variance is reduced for negative covariance. Note that the antithetic method is particularly strong for component-wise monotonic functions f.
4. Explain why it is practical to simulate $\log S_t$ instead of S_t. (Hint 1: Can S_t take negative values in practice? Hint 2: What advantages does the stochastic differential equation for $\log S_t$ in the Black-Scholes model offer over the differential equation for S_t?)
5. Show that for $s_1 < s_2 < \cdots < s_k$ it holds that

$$W_s \big| W_{s_1} = x_1, W_{s_2} = x_2, \dots, W_{s_k} = x_k$$
$$\sim N\left(\frac{(s_{i+1}-s)x_i + (s-s_i)x_{i+1}}{s_{i+1}-s_i}, \frac{(s_{i+1}-s)(s-s_i)}{s_{i+1}-s_i}\right),$$

where $s_i < s < s_{i+1}$ are the immediate neighbors of s.

Exercises with Mathematica

5. Initialize Mathematica's pseudo-random generator `RandomReal` and apply it to numerically compute the integral

$$\int_{[0,1]^5} x_1 x_2^2 (x_3 - x_4 x_5)\, dx_1 dx_2 dx_3 dx_4 dx_5.$$

Determine a suitable choice of N empirically and compare your result to the exact solution of the integral.
6. Use Mathematica to generate a sample of 1,000 points from an Exp(0.5) distribution. Compute the sample mean to give an estimate of the mean of the distribution. How far is the simulated off the theoretical mean? Give an asymptotic 95% confidence interval for the mean, based on your simulation.
7. Explain the functioning of the following Mathematica code.

```
T = 1; Δt = 1/250; m = T/Δt; ρ = -0.8; t
= AbsoluteTime[]; W = {{0,0}}; For[i = 1, i <= m, i++,
  Z1 = RandomReal[NormalDistribution[0,√Δt]];
  Z2 = ρ*Z1 + √(1 - ρ²)
```

```
        *RandomReal[NormalDistribution[0,√Δt]];
    W = Append[W,{W[[i,1]]+Z1, W[[i,2]]+Z2}]];
        AbsoluteTime[]-t
```

8. Use Euler's scheme (5,000 paths and $\Delta t = 1/100$), to numerically price a European call option with maturity $T = 1$ year and strike price $K = 110$ in the Black-Scholes model with $S_0 = 100$, $r = 0.04$ and $\sigma = 0.2$. To do so, simulate $\log S_t$ and compare the attained option price with the result from the explicit Black-Scholes formula. Compute the confidence interval and finally double the number of simulated paths. How does the confidence interval change? What happens if you set $\Delta t = 1$? Comment on the running time of the method.
9. Use the same set of parameters as in Exercise 8 and compare the speed and accuracy of Euler's scheme with the explicit difference method of Chapter 10 applied to the Black-Scholes equation (7.1).
10. Modify the code of Exercise 7 such that the risk-neutral log-price process in the Heston model with parameters $S_0 = 1$, $v_0 = 0.0654$, $r = 0.03$, $\kappa = 0.6067$, $\theta = 0.0707$, $\lambda = 0.2928$, $\rho = -0.7571$ can be simulated with Euler's scheme (use step size $\Delta t = 1/25$ and maturity $T = 1$). Hereby, set the variance to 0 if $v_i < 0$ for some i (where the variance would become negative due to the discretization).
11. Importance sampling can help to increase the efficiency of simulation for the pricing of options. For example, in the case of a knock-in barrier option where the barrier lies far away from the current stock prices, most simulated paths would naturally return a pay-off of 0. However, if the drift of the process is adapted, such that many paths will hit the barrier (and the resulting pay-offs are transformed as per the scheme), one attains significantly lower variance. Use the following Mathematica code to price a knock-in barrier option with knock-in level $U = 140$, strike price $K = 110$ and discrete observation times $i/16$, $i = 1, \ldots, 16$ (which corresponds to 4 observations per year) in the Black-Scholes model of Exercise 8.

```
T = 1; n = 5000; Δt = 1/16; m = T/Δt; μ =
0.4; K = 110; U = 140; t = AbsoluteTime[];
Payoff = {}; For[j = 1, j <= n, j++,
  S = {Log[100]}; w = 1;
  For[i = 1, i <= m, i++,
    Z = RandomReal[NormalDistribution[μ-σ^2/2, σ√Δt]];
    w *= Exp[((Z-Δt*(μ-σ^2/2))^2
             -(Z-Δt*(r-σ^2/2))^2)/(2*Δt*σ^2)];
    S = Append[S, S[[i]]+Z];];
  Payoff = Append[Payoff, If[Exp[Max[S]]>=U,
    w*Exp[-r*T]*Max[Exp[Last[S]]-K, 0], 0]];]
AbsoluteTime[]-t
```

Test how many simulation runs are required, if importance sampling is not used, in order to obtain a confidence interval of equal size as above. How do the runtimes of the two methods compare? How does the performance change based on changes in the parameter μ? What conclusions would you draw from this?
12. Use MC simulation (10,000 sample paths) to price an Asian call option in the Black-Scholes model with maturity $T = 1$ and monthly averaging, $S_0 = K = 100$, $r = 0.05$ and $\sigma = 0.2$. Determine the size of the confidence interval. Test the level of variance reduction, if a European call option is chosen as control variate, and determine the according confidence interval. What is the resulting correlation? (Hint: it will be of computational advantage here to use the same sample paths for the estimation of the Asian and the European option prices, so that a control variate with high correlation is generated).
13. Use the Heston model of Exercise 10 to price a European call option with strike price $K = 1$. Generate 5,000 sample paths and base your simulation on (a) the Euler scheme and (b) the Milstein scheme.

14. Construct the first 1,000 points of a 4-dimensional Halton sequence with bases $b_1 = 2, b_2 = 3, b_3 = 5, b_4 = 7$ by writing a simple Mathematica code.
15. Plot the first 200 points of a Halton sequence in dimensions 21 and 22. Is the discrepancy of the resulting sequence on $[0, 1]^2$ still satisfying?
16. In Exercise 11, use a 16-dimensional Halton sequence with the bases being the first 16 prime numbers instead of the random numbers.
17. Use the Mathematica command

```
BlockRandom[SeedRandom[
    Method -> {"MKL", Method -> {"Sobol",
        "Dimension" -> k}}];
  RandomReal[1, {n, k}]]
```

to generate a k-dimensional Sobol sequence of length n. Use the sequence instead of the random numbers in Exercise 11. How does this affect the result?
18. Solve Exercise 17 by a Brownian bridge construction and comment on how the results compare.

Calibrating Models – Inverse Problems 12

In the previous chapters we studied several model choices to describe stock price and interest rate dynamics. When using models to valuate derivatives or to obtain a hedging strategy, the used parameters will greatly impact the results. While there is broad agreement of how to model many problems in physics (such as the thermal conductivity of copper at room temperature), financial markets are fundamentally different. Many market participants have different views on the distributions of market variables, and market prices of liquid assets only represent an economic equilibrium resulting from those different views. In relation to stock options, note that quoting the implied Black-Scholes volatility of a specific (strike, maturity) option is simply a different way of stating a price. The actual volatility and the dynamics at which the stock price will move are therefore initially not related to (quoted) implied volatilities.

A typical procedure for valuing a derivative can be described as follows.

1. Choose a model (or several models, if *model risk* should be assessed).
2. Calibrate the model input parameters to the market prices of liquid instruments.
3. Compute the value of the derivative in the calibrated model and, if required, value sensitivities (typically by numerical methods).

So far, we have extensively dealt with the first and the last item, i.e. the selection of a model and the valuation of derivatives in a chosen framework. We will now focus on the calibration of the model parameters, such as $\sigma(S,t)$ in the Dupire model, the parameters of the Heston model, or the parameter functions in the Hull-White or Black-Karasinski model. All these calibrations are examples of *inverse problems*, i.e. one looks at observed or desired outcomes (in this case market prices of liquid instruments) and aims to determine the causes (parameter functions of the models) by employing appropriate quantitative methods.

Inverse problems can pose severe stability challenges. To illustrate this, consider a very simple inverse problem in financial mathematics. Assume that we observe the quoted prices $Z(0,T_i)$ of zero-coupon bonds that pay 1 at maturity T_i, and try to determine the forward short rates $r(\tau)$ $(0 \le \tau \le \max_i(T_i))$ that explain the prices as

H. Albrecher et al., *Introduction to Quantitative Methods for Financial Markets*,
Compact Textbooks in Mathematics, DOI 10.1007/978-3-0348-0519-3_12,

$$e^{-\int_0^T r(\tau)d\tau} = Z(0,T).$$

Defining $y(T) := -\ln Z(0,T)$ implies for sufficiently smooth y that

$$r(t) = \frac{d}{dt}y(t).$$

Typically $Z(0,T)$ will not be stated as a single price, but as a price interval (as *bid-ask spread*) implying that there is some (small) uncertainty on the exact bond price. Assume $y(t)$ and $r(t)$, respectively, to be given through the exact prices and define $y_n^\delta(t) := y(t) + \delta \sin(nt/\delta)$ to be the bond yields (multiplied by maturity) used for calibration. Then $r_n^\delta(t) = dy_n^\delta(t)/dt = r(t) + n\cos(nt/\delta)$ is the calibrated short rate. It follows that $\max_t |y_n^\delta(t) - y(t)| = \delta$ for all n, but $\max_t |r_n^\delta(t) - r(t)| = n$. One can therefore see for large n that small fluctuations in the zero-coupon bond curve $y(t)$ can lead to arbitrarily large changes in the resulting forward short rates $r(t)$ due to differentiation (see Exercise 2).

12.1 Fitting Yield Curves in the Hull-White Model

The problem of determining the parameter function a in a one-factor Hull-White model (see Section 9.3) is relatively complex. Assume that the reversion speed b and the volatility σ are known, and so is the starting value r_{t_0} of the short rate. Recalling the formulas (9.7) and (9.8), observing the bond prices for all T implies the knowledge of the values $\alpha(t_0, T)$, so that, to find a, one has to solve

$$\int_{t_0}^{T} a(s)\,(e^{-b(T-s)} - 1)\,ds = g(T) \qquad (12.1)$$

where g results from algebraic manipulations of (9.8). This is a linear integral equation of the first kind, and the sought-for function a only appears inside the integral.[1] The integral operator which maps the unknown function a to g in our example damps oscillations in a,[2] and the damping effect becomes stronger with higher frequency of the input a. Conversely, if g is distorted by noise due to observation errors, a naïve solution procedure that applies discretization to the system of the integral equation will increasingly magnify this error, as the perturbation frequency goes up.[3]

[1] Integral equations have been studied extensively, see e.g. Engl [27].

[2] This is a general property of integral operators with bounded integration region and which satisfy weak conditions (e.g. quadratic integrability) for the integration kernel (which is an exponential function in our case). Under such conditions, the integral operator is found to be a compact operator whose singular values tend to 0.

[3] We have used exactly this property for the construction of the above example when determining the forward short rates.

Example

Consider a Hull-White model with reversion speed $b = 0.8$, volatility $\sigma = 0.7\,\%$ and a time horizon of 25 years. For $T \in [0, 25]$, the actual interest curve (zero rates with continuous compounding) shall be given by $0.06 - 0.03\exp(-T/4)$. This implies that interest rates start at 3 % at the short end and increase to 6 % after 25 years.

We shall now perturb values at given nodes $(t_1, t_2, ..., t_n)$ of the interest curve with a relative error of at most 0.1%, which refers to a maximum absolute perturbation of 0.006% (or: 0.6bps)[4]. Yearly (low frequency) and monthly (high frequency) discretization then leads to the solutions for a plotted in Figure 12.1 (bold line: exact solution).

Despite being able to base the second calibration on more data points, the results are worse and even unusable.

Well- and Ill-Posed Problems

In 1923 Jacques Hadamard postulated that a problem $Kf = g$ (with given functional K) in physics is *well-posed* if it satisfies the following properties:

1. There exists a solution for arbitrary g.
2. This solution is unique.
3. The solution depends continuously on the data.

If any of the three properties are violated (the third one is the most critical one), one refers to the problem as being *ill-posed.* Integral equations of the first kind are an example of ill-posed problems, and arbitrarily small noise on the right-hand side g can lead to large changes in the solution f. A classical way of stabilizing ill-posed problems with noisy right-hand sides g_δ (so that g deviates from g_δ by a maximum of δ) is *Tikhonov regularization*. Specifically, instead of $Kf = g_\delta$, one solves

$$f_\delta^\alpha = \arg\min \left(\|Kf - g_\delta\|_2^2 + \alpha_\delta \cdot \mathcal{P}(f) \right) \tag{12.2}$$

for appropriate α_δ, with $\|\cdot\|_2$ denoting the L_2 norm and $\mathcal{P}(f)$ being a penalty term (e.g. the squared L_2 norm of f or of a derivative of f).[5]

Consider the simple case where $\mathcal{P}(f) = \|f\|_2^2$. For $\delta \to 0$ and $\delta^2/\alpha_\delta \to 0$, the f_δ^α's will converge to the least-square–minimum-norm solution of $Kf = g$, which is the solution with the minimum penalty term among all solution candidates for which the residual $\|Kf - g\|_2$ is minimal.[6]

---

[4] 1 bp = 1 basis point = 0.01 %.

[5] This penalty term can contain a priori information on a guess for the true function f^*, for example, by measuring the distance to f^*.

[6] The proof of this statements requires some profound techniques of Functional Analysis (see Engl, Hanke & Neubauer [28]). In this case the optimization problem (12.2) (in the infinite-dimensional setting) will be equivalent to the solution of $(K^*K + \alpha I) f_\delta^\alpha = K^* g_\delta$, where K^* is the operator adjoint to K and I the identity operator.

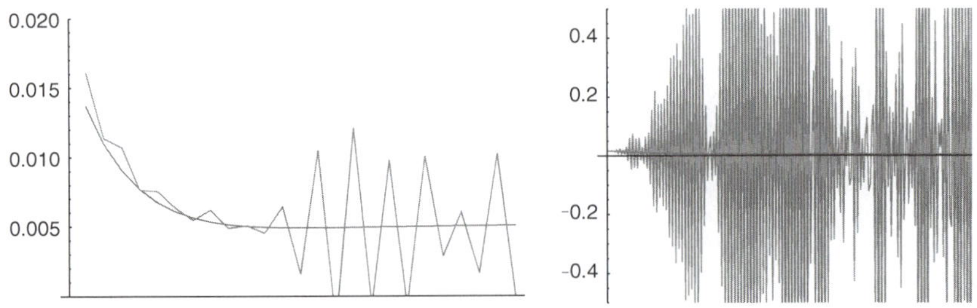

Fig. 12.1 Hull-White parameter a for yearly (left) and monthly (right) data (both the exact solution and the solution for the perturbed data)

The Finite-Dimensional Case: Bad Condition Numbers

In practice one can only obtain a finite number of reference points (nodes) for the fitting of curves, which then requires solving a system of linear equations (e.g. when determining the yield curve based on finitely many reference points). One will often find the system matrices to be ill-conditioned (see Figure 12.1), so that regularization techniques are required as the right side g will only be known up to a certain level of accuracy (e.g. 4 or 6 decimal places).

The condition numbers of the system matrices will depend on the width of the time discretization grid. Typically the instability issue becomes more severe as the time grid becomes narrower. In the finite-dimensional case, one solves the regularized normal equation $(M^{\mathrm{T}}M + \alpha I)x = M^{\mathrm{T}}y$ instead of the ill-conditioned equation $Mx = y$ (with system matrix M).

Reversion Speed and Volatility

Up to now we have assumed that the reversion speed and the volatility are known, and we have calibrated the parameter a of the Hull-White model to the interest curve. To identify the remaining two parameter functions, one requires market data for instruments that are more sensitive to the stochastic behavior of interest rates. Among the liquid instruments, caps and swaptions are obvious candidates.

Experience shows that calibrating the reversion speed and the Hull-White volatility based on only swaptions works well, while performing the calibration solely based on cap prices does generally not give satisfactory results (in the case of caps, the single caplet layers are separated from each other in a way that they provide too little information on the speed of reversion).

Let $\mathbf{f} = (f_1, f_2, \ldots)$ be a vector collecting the different model parameters. The typical procedure is now to compute model prices $V_j(\mathbf{f})$ (for example, of traded caps or swaptions) and to determine $\mathbf{f}$ so that the model prices $V_j(\mathbf{f})$ are as close to the market prices P_j as possible, i.e. we solve

$$\min_{\mathbf{f}} \sum_j |V_j(\mathbf{f}) - P_j|^2.$$

Note that this minimization problem requires the ability of evaluating the functions efficiently for arbitrary combinations of parameters. Being able to efficiently compute gradients for the target functions with respect to the parameters will be an advantage. Note that the choice of writing the above error functional in terms of squared price deviations is rather arbitrary. As an alternative, one could also express market prices as Black76 volatilities (which will frequently be quoted) and run the optimization on the volatilities instead of the prices themselves.

In that case, one has to pay attention to the fact that for far in-the-money or out-of-the-money options, the option prices will hardly react to changes in the Black76 volatility, so that this approach will be more suitable for at-the-money options (where the strike is close to the forward rate).

If the reversion speed and volatility are found to depend on time (for example, as piecewise constant functions), rather than being constant, the danger of oscillation increases greatly. This case would normally result in a non-linear problem that should be treated by appropriate regularization methods for nonlinear problems.

12.2 Calibrating the Black-Karasinski Model

Identifying the parameters in the Black-Karasinski model does not require many additional techniques from a theoretical viewpoint. In practice, however, calibration is considerably more cumbersome, as no analytical formulas are available for the model prices of instruments (such as bonds, caps or swaptions). The prices and their respective first derivatives with respect to the model parameters have to be determined by numerical methods. Approximating the first derivatives by difference quotients would make the evaluation of functions computationally time-consuming. It is established practice in this case to use so-called *adjoint methods*, for which the evaluation of the gradient only requires the same computational effort as the evaluation of the function itself. Markets in which short-term interest rates are very low (as observed for e.g. the CHF, EUR or JPY in recent years) pose a major challenge when calibrating the Black-Karasinski model. Low values of r require significant volatilities in order to produce reasonable fluctuations. For higher values r, on the other hand, high volatilities could be potentially dangerous. For a meaningful calibration of the model, it can then be practical to artificially inflate short-term interest rates.

12.3 Local Volatility and the Dupire Model

The local volatility model (see Section 8.2) uses a two-dimensional function $\sigma(S, t)$ as volatility parameter. We can now see the price of a call as a function of its strike price K and its maturity T. Under the assumption that the risk-free interest rate r

only depends on time, Dupire was able to prove that, for fixed t_0 and known stock price S_{t_0}, the call price $C(S, K, t, T)$ satisfies both (8.3) and its *dual* condition[7]

$$-\frac{\partial C}{\partial T} + \frac{\sigma^2(K,T)\,K^2}{2}\frac{\partial^2 C}{\partial K^2} - r\,K\,\frac{\partial C}{\partial K} = 0. \tag{12.3}$$

Rearranging this equation gives the function $\sigma(K,T)$ describing the *volatility surface*,

$$\sigma(K,T) = \sqrt{\frac{\frac{\partial C}{\partial T} + rK\frac{\partial C}{\partial K}}{\frac{K^2}{2}\frac{\partial^2 C}{\partial K^2}}}. \tag{12.4}$$

If call prices $C(K,T)$ were available in the market for arbitrary (K,T) combinations, one could simply use (12.4) to determine the entire surface $\sigma(K,T)$. In practice, however, $C(K,T)$ will only be observable for certain (traded) nodes (K,T). Direct interpolation of option prices from the given reference points (with subsequent calculation of $\sigma(K,T)$ by (12.4)) is generally not advisable. Differentiating is itself numerically instable, and if the strike prices significantly deviate from the current stock price, the second derivative $\partial^2 C/\partial K^2$ is close to 0, so that dividing by this small number in (12.4) can greatly magnify errors.

In practice, market data of the implied (Black-Scholes) volatilities will be given as depicted in Figure 8.2. It is then common practice to calibrate the Dupire model either by generating a smooth implied volatility surface (e.g. using splines or other basis functions) and applying a variant of the Dupire formula directly using implied volatilities (and derivatives thereof) or as follows:

1. Translate the implied market volatilities to market option prices.
2. Choose a finite-dimensional ansatz for the unknown Dupire volatility $\sigma(S,t)$ (for example, piecewise constant or piecewise linear) and some initial function $\sigma^0(S,t)$ (set $k = 0$).
3. Use the current $\sigma^k(S,t)$ to calculate the Dupire prices at the nodes and compute a Tikhonov functional (see Section 12.1), consisting of an error functional (e.g. the sum of the squared errors) and a regularizing term (which could e.g. penalize oscillations in $\sigma^k(S,t)$).
4. Apply some updating algorithm to reduce the value of the Tikhonov functional. This produces the new (improved) volatility $\sigma^{k+1}(S,t)$.
5. If the result is not satisfying, increase k by 1 and go to step 3.

[7]The Fokker-Planck equation, which describes the time evolution of the probability density function of the transition distribution of the stock price under the risk-neutral measure, offers one possible way of deriving this dual equation. Hereby one takes a Dirac-delta distribution as starting distribution. Under the risk-neutral measure the local volatility function is uniquely determined by the call prices (for all strike/maturity pairs).

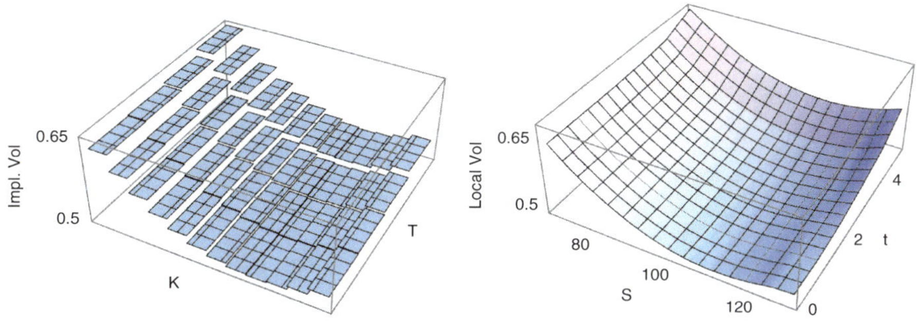

Fig. 12.2 Left: synthetic volatility data (as implied by call-prices, strikes: 70 to 130 %, maturities 1 to 5 years). Right: volatility surface $\sigma(S,t)$ calibrated to the data points

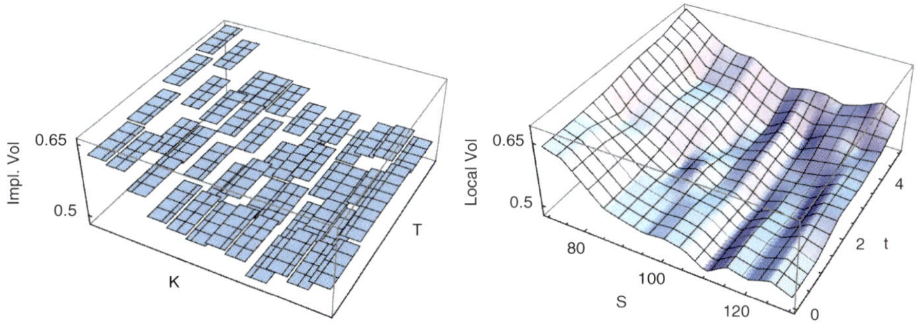

Fig. 12.3 As in Figure 12.2, data (left) is distorted by noise by up to two percentage points

Numerically this can prove to be a complex procedure. Each iteration step requires the numerical solution of a differential equation (to obtain the Dupire option prices from the volatilities), as well as finding a 'better' volatility $\sigma^{k+1}(S,t)$. Efficient updating algorithms, such as the quasi-Newton method, require the determination of the gradient of the to-be-minimized function. In this case, one has to evaluate the derivative of the Tikhonov functional with respect to σ at $\sigma = \sigma^k$. Adjoint methods often turn out to be efficient for the computation of the gradient (cf. Section 12.2), so that the computational complexity of finding the gradient is again of the same order as evaluating the functional itself. It is a great advantage that this method also works if the implied volatility data points are distorted by noise (see Figures 12.2 and 12.3).[8]

[8]Note that market data will always be noisy due to bid-ask spreads.

12.4 Calibrating the Heston Model or the LIBOR-Market Model

When calibrating the Heston model, one has to determine its five parameters $\kappa, \lambda, \rho, \theta, v_0$ (with initial variance v_0). To calculate the residual, one has to compute the errors between the Heston prices of the options and the corresponding market prices. Although this could be expected to be a relatively simple problem due to the low number of model parameters, this is actually not the case: the objective functional (for example, the sum of the squared absolute or relative errors) will typically produce a large number of local minima, so that it will be of advantage to combine techniques of local and global optimization (e.g. the simulated annealing method). Alternatively it can be constructive to use a large number (> 100) of initial values for local optimization algorithms and to choose the 'best' minimum. The initial values could, for instance, be determined by a low-discrepancy sequence (cf. Section 11.2). Moreover one should keep in mind that it will be desirable for the obtained parameters to satisfy Feller's condition. This can be achieved by applying penalty terms at the bounds of the feasible region.

Calibrating the parameters of the Libor market model leads to similar challenges as in the Heston model. Again, one will obtain a large number of local extrema, and optimization algorithms that allow for many automated re-starts appear promising.

12.5 Key Takeaways, References and Exercises

After working through this chapter you should understand and be able to explain the following terms and concepts:

- Inverse problems can be very complex computationally, higher frequency data points can lead to unusable results if model calibration is implemented naïvely
- Well- and ill-posed problems: calibration problems in finance are often ill-posed
- The Tikhonov regularization: idea, resulting numerical problem
- Calibration issues in the Dupire, Heston and Libor-Market model

References

Regularization methods for inverse problems and their convergence properties are discussed in Engl, Hanke & Neubauer [28] based on functional analysis techniques. More recent methods (in particular, iterative) for nonlinear inverse problems can be found in Kaltenbacher, Neubauer & Scherzer [44]. Egger & Engl [25] provide an analysis of convergence for the identification of local volatility. For more background on the simulated annealing method, consult Brémaud [12]. Finally, [29] is an easily accessible article on inverse problems in financial mathematics, while Cont & Tankov [18] also cover inverse problems in the context of Lévy processes.

Exercises

1. For fitting a yield curve in the Hull-White model, assume that the speed of reversion and the volatility are given and constant (to keep things simple). Solve the integral equation with respect to $a(\cdot)$ by differentiating appropriately.

Exercises with Mathematics and UnRisk

2. Explain the functionality of the following Mathematica code.

```
RandomNormal[a_,b_]:=Random[Normal
    Distribution[a,b]];
NodesPerYear=10;

BondValues= Table[{i/NodesPerYear,
          Exp[-(0.05+RandomNormal[0,0.0001])*i/Nodes
              PerYear]},
          {i,0,10*NodesPerYear}];

ListPlot[BondValues,PlotJoined -> True]

f[t1_]:=-Log[Interpolation[BondValues][t1]]
Plot[Derivative[1][f][x],{x,0,10}]
```

   What happens if `NodesPerYear` is 2? Or 100?
3. Use the UnRisk command `CalibrateLocalEquityVolatility` to calibrate some volatility surfaces.

Case Studies: Exotic Derivatives 13

Today's financial markets offer a wide range of complex financial products. In this chapter we will introduce several structured financial instruments and discuss ideas for their valuation. The exercises at the end of the chapter will then further illustrate the specific features of the presented instruments.

13.1 Barrier Options and (Reverse) Convertibles

Convertible bonds are bonds that typically pay a relatively low running coupon but give the investor the option to exchange the bond for a fixed number of stocks at maturity T. The bond investor therefore has a call option on some underlying stocks.[1] Convertible bonds are also popular with growth companies (e.g. start-ups) that cannot afford to initially pay high cash coupons on their bonds but are willing to give up equity upside in the future to the bond investors in case the company develops successfully. *Contingent convertibles* (CoCos), on the other hand, are bonds that are automatically converted into equity if some defined trigger event occurs. CoCos have been much discussed as instruments with great potential to mitigate the adverse effects of market downturns on the financial sector. For example, banks or insurance companies could issue CoCos during strong markets, so that equity (or: capital) losses during weak markets could be covered by converting CoCo bonds to equity. CoCos are structurally similar to *reverse convertibles* which pay a higher coupon, as the conversion option is now held by

[1]French telecom companies (e.g. France Telecom) issued a large volume of convertible bonds around the year 2000. Sometimes one distinguishes between convertible bonds and *exchangeable bonds*. While convertibles are issued by the company that offers to exchange the bonds against its own stocks in the future, exchangeable bonds are typically issued by third parties that would like to sell stocks in a venture in the future (e.g. for planned privatization of companies owned by public entities)

H. Albrecher et al., *Introduction to Quantitative Methods for Financial Markets*, Compact Textbooks in Mathematics, DOI 10.1007/978-3-0348-0519-3_13,

the bond issuer. The investor effectively has a short position in a put option on some underlying stocks.

Knock-in derivatives are instruments for which an underlying derivative becomes only effective once a certain (observable) *trigger event* occurs.

Example (Knock-in reverse convertible)

In March 2005, ABN AMRO issued the following instrument with Apple Inc. stocks as underlyings:
Issuance/maturity date: 16th of March 2005/16th of March 2006
Face Amount: 1,000 USD
Apple Inc. stock price on 11th of March 2005 (closing price): 40.27 USD
Knock-in level (for trigger event): 28.19 USD (=70 % of 40.27)
Coupon: 11.25 % p.a, semi-annual payments

In any case, the investor will receive a coupon of $11.25\,\% \cdot 0.5 \cdot 1000 = 56.25$ on 16/09/2005 and on 16/03/2006. If Apple's stock price has not dropped below 28.19 USD by maturity, the reverse convertible will simply repay the principal at $100\,\% = 1{,}000$ USD on 16/03/2006. If Apple Inc. trades below the knock-in level at least once prior to maturity, the bond issuer can choose to repay either the principal of $1,000$ USD or to deliver 24.832 Apple stocks at maturity T. The issuer will only exercise this (contingent) conversion right if Apple Inc. trades below $1{,}000/24.832 = 40.27$ USD at maturity.

From the investor's point of view, the reverse convertible can be divided into two components: (a) a bond with a coupon of 11.25 % and (b) the obligation (but not the right) to purchase 24.832 Apple stocks at a total price of 1,000 USD contingent on Apple's stock price hitting 28.19 USD at some time up 16/03/2006. Neglecting coupon payments, this implies a pay-off P on 16/03/2006 of

$$P = 1{,}000 - 24.832(40.27 - S_T)^+ \, 1_{\{\min_{t_0 < t \leq T} S_t \leq 28.19\}}, \tag{13.1}$$

with t_0 being the initial date (here: 16/03/2005), T the maturity (here: 16/03/2006) and S_t is Apple's stock price at time t ($1_{\{\}}$ denotes the indicator function).

Ex-post observation: the Apple stock did not hit the barrier during the life of the instrument, so that the return to the investors was simply given by the 11.25 % p.a. coupon payments. As a point of reference, a direct investment in Apple stocks produced a return of 60 % over the same year, partly due to the success of Apple's iPod.[2]

[2] 'It is difficult to make predictions, especially about the future.' (accredited to Mark Twain)

Knock-Out Options and Knock-In Options

For the pay-off (13.1) only the term $24.832(40.27 - S_T)^+ 1_{\{\min_{t_0 < t \le T} S_t \le 28.19\}}$ is stochastic (i.e. depends on Apple's stock price). An option offering such a pay-off structure is referred to as *knock-in (put) option*, as the option only becomes effective if the stock price drops low.

Let us assume that this put option on Apple stocks is effective from the beginning, but would be canceled (or: *knocked-out*) once the stock price hits the value 28.19 USD. The knock-out put option would then have the pay-off $24.832(40.27 - S_T)^+ 1_{\{\min_{t_0 < t \le T} S_t > 28.19\}}$. At any time up to expiry exactly one of the knock-in and the knock-out option is alive. The sum of the values of these two options must therefore correspond to the value of a plain vanilla put option (in case of European barrier options). Knock-out options are easier to handle analytically, and they can be priced in the Black-Scholes model based on the distribution of the *first-passage time* of the Brownian motion.

Theorem 13.1 (First-passage time of the Brownian motion). *Let $X_t = X_0 + \mu t + \sigma W_t$ with $X_0 \ge 0$, $\mu \in \mathbb{R}$, $\sigma > 0$ and W_t a Brownian motion. The first-passage time τ_h for $h < 0$ is then defined as $\tau_h = \inf\{t > 0 : X_t = h\}$. The cumulative distribution function of τ_h is given by*

$$\mathbb{P}[\tau_h \le t] = \Phi\left(\frac{-X_0 - \mu t}{\sigma\sqrt{t}}\right) + e^{-2X_0\mu/\sigma^2}\Phi\left(\frac{-X_0 + \mu t}{\sigma\sqrt{t}}\right) \quad (13.2)$$

We will now prove this theorem for the special case $X_0 = 0$, $\mu = 0$ and $\sigma = 1$ (which is also known as *reflection principle*).[3]

Example (Reflection principle)

We consider the Brownian motion W_t and aim to determine its probability of not crossing some value $h < 0$ up to time T. Define $\tau_h = \inf\{t > 0 : W_t = h\}$ and the corresponding process that is *reflected* at h,

$$\widetilde{W}_t = \begin{cases} W_t & \text{for } t < \tau_h, \\ h - (W_t - h) & \text{for } t \ge \tau_h. \end{cases}$$

By definition, $W_t - W_{\tau_h}$ is independent of W_{τ_h} and normally distributed with mean 0.[4] It follows that

[3]The proof of the theorem for arbitrary μ and σ builds on arguments of stochastic analysis which are beyond the scope of this text. For details, see e.g. Karatzas & Shreve [45].

[4]This intuitive property is a consequence of the *strong Markov property* of the Brownian motion, and a mathematically rigorous proof is given in the theory of stochastic processes.

$$\mathbb{P}[\tau_h \le t, \widetilde{W}_t < x] = \mathbb{P}[\tau_h \le t, W_t < x]$$

and

$$\mathbb{P}[\tau_h \le t, W_t > h] = \mathbb{P}[\tau_h \le t, W_t < h].$$

$W_t < h$ for some $t < T$ implies that h has been crossed and, hence, $\mathbb{P}[\tau_h \le T, W_T < h] = \mathbb{P}[W_T < h]$. The probability of a standard Brownian motion hitting the barrier h on the interval $[0, T]$ is

$$\begin{aligned}\mathbb{P}[\min_{[0,T]} W_t \le h] &= \mathbb{P}[\tau_h \le T] = \mathbb{P}[\tau_h \le T, W_T > h] + \mathbb{P}[\tau_h \le T, W_T < h] \\ &= 2\,\mathbb{P}[\tau_h \le T, W_T < h] = 2\,\mathbb{P}[W_T < h] = 2\,\Phi(-h/\sqrt{T}),\end{aligned}$$

which corresponds to (13.2) for $\mu = 0$ and $\sigma = 1$.

Knowing the distribution of the first-passage time can be applied to price barrier options and similar financial products. We will return to this result at a later point, and Exercises 1 and 3–5 will discuss additional aspects of barrier options.

13.2 Bermudan Bonds – To Call or Not To Call?

Section 9.1 introduced European swaptions which give the right to become the counterparty in a payer or receiver swap at some maturity T. At time T it is straightforward to decide whether to exercise the option: in the case of an underlying payer swap, the option will be exercised only if the then applicable market swap rate lies *above* the fixed rate as agreed in the swaption contract. In practice, bond or loan contracts will often include *pre-payment options* for the borrower.[5] Bonds with options to pre-pay in full are referred to as *callable bonds*. The following example analyzes the call decisions of a bond issuer based on a one-factor short-rate model.

Example (Bermudan callable fixed-rate bond)

Issuance/maturity date: 17th of July 2012/17th of July 2023
Coupon: 5.5 % p.a., paid annually
Face amount: 100
Starting in 5 years (i.e. from 2017), the issuer has the right to prepay the principal (or: to call the bond) in full on coupon days.

Assume that the bond has not been called up to 2022. The issuer then decides on the 2022 coupon day whether to call the bond, i.e. whether to immediately repay the

[5] A *pre-payment option* is the right of the borrower to repay at least parts of the (loan) principal early. If the possibility to pre-pay is restricted to a discrete number of pre-payment dates, the right (option) will be called *Bermudan*.

face amount of 100, rather than 105.5 (principal repayment plus one year's interest) one year later. In principle, this call decision should be solely based on the level of the short-rate r on the 2022 coupon date (and the resulting one-year interest rate). The value $V(r, 2022)$ of the bond to the investor on the 2022 coupon day now depends on whether the call is exercised, and we write $V(r, 2022) = 5.5 + V_{\text{call}}(r, 2022) = 105.5$ if the bond is called, and $V(r, 2022) = 5.5 + V_{\text{keep}}(r, 2022)$ if it is not (V_{keep} denotes the value of the bond if it is repaid in 2023). More generally, we find for (Bermudan) call dates t_B that

$$V(r, t_B) = \text{Coupon} + \min\Big(V_{\text{call}}(r, t_B), V_{\text{keep}}(r, t_B)\Big).$$

In between two call dates t_{B_i} and $t_{B_{i+1}}$, V_{keep} satisfies the respective differential equation of the chosen short-rate model with boundary condition

$$V_{\text{keep}}(r, t_{B_{i+1}}) = V(r, t_{B_{i+1}}),$$

so that the product can be priced recursively, from maturity T down to today. Exercises 2 and 6 will further illustrate Bermudan callable bonds.

13.3 Bermudan Callable Snowball Floaters

In times of low interest rates it is popular to issue *snowball instruments*.[6] Consider the following example.

Example (Snowball floater)

Term to maturity: 10 years, coupon: paid semi-annually (i.e. 20 coupons)
Nominal amount: 100
Coupon(1) = Coupon(2) = 5.5 % p.a.
$\text{Coupon}(i+1) = \max\Big(0\,\%, \text{Coupon}(i) + \text{Step}(i) - \text{Libor6M}(t_i)\Big)$ for $i = 2, \ldots, 19$.
$\text{Step}(i) = 3\,\% + i \cdot 0.1\,\%$.

The reference interest rate (here: Libor6M) for the computation of the coupon amounts will be set either at the beginning or, more commonly, at the end (in arrears) of the respective coupon period. In times of low interest rates (e.g. in the periods 2004–2005 or 2009–2012) a snowball floater can appear attractive to a buyer due to its relatively high coupon payments.

Even in the simplest case of a non-callable snowball floater, it is not possible to statically replicate the coupons by fixed or floating (Libor) coupons due to the floor

[6]The term 'snowball' is based on the image that a snow ball (as used when building a snowman) becomes increasingly bigger as it is rolled in the snow.

at 0%. In practice, snowball floaters will typically provide call rights to the issuer, which further complicates the analysis.

The issuer will exercise its call right if the retention value to the investor (under consideration of possible future calls) is higher than the value if called. As coupons are path-dependent, the optimal strategy on a call date will depend not only on market interest rates, but also on the most recent coupon size. Exercises 7 and 10 will further illustrate snowball floaters.

13.4 More Examples of Exotic Interest Rate Derivatives

In the following we will discuss three more *exotic* instruments whose valuation is implemented in UnRisk.

Steepener Instruments

In the case of snowball floaters, the issuer would typically expect rising market interest rates (e.g. Libor rates) and hence falling coupons. *Steepener instruments* (or: *CMS spread instruments*), on the other hand, focus on the difference in interest rates for different terms.

CMS spreads are commonly based on the difference between the 10-year and the 2-year swap rate (CMS10Y – CMS2Y).[7] If this difference is large (as the interest rates at the short end are relatively low), we will refer to the interest curve as *steep*. A major investment bank issued the following instrument in March 2005:

Example (Steepener)

Term to maturity: 15 years, notional: 1,000 USD
Coupon: paid quarterly, in the first year: 15 % p.a.
Thereafter: max(0, 20 × (CMS30Y − CMS10Y)), with CMS30Y and CMS10Y being the then quoted 30- and 10-year USD swap rates as quoted on a coupon calculation day

Note that steepeners frequently provide call rights to the issuer. As a potential change in shape (or: *rotation*) of the interest curve plays a crucial role when pricing these instruments, one-factor short-rate models will not be suitable. Steepeners and similar instruments will typically be priced using Libor market models (see Exercises 11 and 12).

[7] CMS is short for *constant maturity swap*. Concretely, CMS10Y denotes the 10-year swap rate as quoted by ISDAFIX. The term 'constant maturity' hereby refers to the fact that the swap rates are always quoted for the same term (e.g. 10 years from the quote day for the CMS10Y).

Range Accruals

For *range accrual notes* it is checked daily if some reference interest rate (e.g. Libor or CMS) lies within a defined corridor (e.g. if Euribor6M $\in$ [3.2%, 3.8%]). If the reference rate is within the corridor on n out of N business days over a coupon period, a coupon of $(n/N) \times C$ is paid on the coupon payment date. C can hereby be a fixed rate (which would then be relatively higher than market bond rates) or a reference rate plus a spread (e.g. Libor6M + 80bps).

Range accruals are also traded for terms in excess of one coupon period. For such instruments it is common to define wider corridors for later coupon periods. Range accruals can come with issuer call rights, and there are also structurally similar instruments that use an exchange rate instead of a reference interest rate for the calculation of the coupon amounts.

Target Redemption Notes

Target redemption notes (TARNs) are an example of *auto-callables*. Such notes offer one or more relatively attractive coupons over some initial fixed-interest period, after which the coupons are calculated based on some reference interest rate. Once the sum total of the paid coupons exceeds a defined *target level*, the note matures so that the principal amount is repaid to the investor. The investor will typically prefer early redemption, while the issuer will hope for low borrowing costs, and hence late repayment.

Example (TARN)

Target level: 40 %
Coupons: 9 years at 4 % p.a., thereafter 5 × (CMS10Y − CMS2Y).
The note is repaid in full as soon as the aggregate coupon amount reaches 40 % of the face amount.

A TARN structure as chosen above can be popular when tax reasons (e.g. to qualify as life insurance investment) require a minimum maturity (in the above example: 10 years).

13.5 Model Risk in Interest Rate Models

Chapter 9 discussed the Hull-White, the Black-Karasinski and a basic version of the Libor market model. The question is now how to choose a particular model class when pricing an interest-rate derivative. Consider the following example.

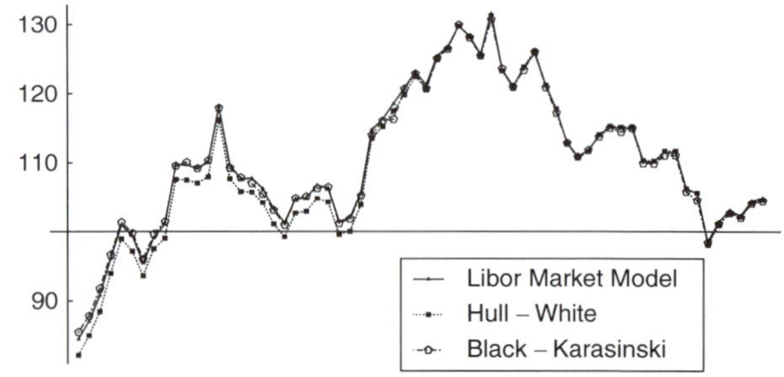

Fig. 13.1 Valuation of a callable reverse floater using different interest-rate models calibrated to market data (2002–2007, valued monthly)

Example (Callable reverse floater 2001–2021)

Issuance/maturity year: 2001/2021, face amount: 100 EUR.
Coupon: $\max(0, 16.5\,\% - 2 \cdot \text{CMS}5Y$ (fixed in arrears), paid annually
Callable: at 100 EUR, annually from 2011.

Calibrating the one-factor Hull-White, the Black-Karasinski and the LMM models to monthly market data (from interest curves, caps and swaptions) over the period 2002–2007, produces the graph in Figure 13.1.

Various models make different assumptions for the distribution of interest rates, so that applying them to the valuations of instruments that depend on the stochastic behavior of interest rates will give different results. In particular, Figure 13.1 shows price differences of up to 3 %. While such deviations appear high at first sight, their size could be considered moderate given the long term to maturity and the leveraged structure of the CMS instrument.

13.6 Equity Basket Instruments

Structured or derivative instruments whose values depend on the development of the price of a set (or: *basket*) of stocks are called *equity basket instruments*. The stocks in the basket could be quoted in the same currency or in different currencies. We now list three examples of such basket products.

Example (Knock-in basket reverse convertible)

Similar to the Apple Inc. example of Section 13.1, a relatively high coupon is paid as long as the knock-in event has not occurred. The difference is now that one deals with a basket of underlyings (rather than a single underlying) and each underlying can trigger the knock-in event. One could now model the individual stocks in the basket by the Black-Scholes dynamics

$$dS_i = \mu_i S_i dt + \sigma_i S dW_i, \quad i = 1, \dots, N.$$

Assume that the stock prices (and the corresponding Brownian motions W_i) are correlated with correlation coefficients ρ_{ij} (and $\rho_{i,i} = 1$), so that $\text{Cov}(dW_i, dW_j) = \rho_{ij}\, dt$. Deriving the price of a European option whose pay-off depends on a basket of stocks under the risk-neutral measure then requires a multi-dimensional version of Itô's Lemma (cf. Chapter 8). The Black-Scholes differential equation for N underlyings that are quoted in the same currency (with risk-free interest rate r) reads

$$\frac{\partial V}{\partial t} + \frac{1}{2}\sum_{i=1}^{N}\sum_{j=1}^{N}\sigma_i\sigma_j\rho_{i,j}S_iS_j\frac{\partial^2 V}{\partial S_i\partial S_j} + \sum_{i=1}^{N} rS_i\frac{\partial V}{\partial S_i} - rV = 0,$$

and the boundary conditions have to be chosen such that they reflect the specific pay-off structure of the derivative.

Example (Altiplano notes)

Altiplano notes pay a relatively high running coupon, but only as long as *all* stocks in a basket price above a certain level (e.g. 70 % of the initial price). There are Altiplano structures that pay missed coupons at a later time once stock prices have recovered. A modified version of such notes might exclude stocks from the basket upon exceptionally bad performance (e.g. if a stock price drops below 50 % of its initial level).

Example (Swing notes)

The coupon of a swing note is calculated as the minimum positive return of the stocks in a defined basket. For example, if the prices of the single stocks in a basket changed by (order by size)..., −5 %, −2.2 %, 1.7 %, 6 %, ..., a coupon of 1.7 % would be paid. A modified version sets the floor of a coupon at the coupon size of the previous period, so that coupon amounts only increase over time ('lock-in').

13.7 Key Takeaways, References and Exercises

After working through this chapter you should understand and be able to explain the following terms and concepts:

- Barrier options, (reverse) convertibles: knock-in vs. knock-out, the link between the first-passage time of an underlying and knock-out options, the distribution of the first-passage time of the Brownian motion
- Bermudan bonds, Bermudan callable snowball floaters: call feature, snowball coupon sizes, basic pricing ideas
- Steepener (CMS) instruments (why are one-factor models inadequate for pricing?), Range accruals, TARNs
- Different interest-rate models will valuate interest-rate products differently
- Basket products: structure of knock-in basket reverse convertible, Altiplano notes, Swing notes

Exercises

The examples stated in this chapter are based on actual products traded in the market. Different market environments will favor different products, since the needs of market participants and regulatory requirements change over time. UnRisk documentation lists the exotic instruments for which pricing tools have been implemented, as well as the mathematical methods these tools are based on. Stochastic methods for the pricing of exotic options are extensively discussed in Dana & Jeanblanc [19].

Exercises

1. Explain why the equation 'Knock-In + Knock-Out = Plain Vanilla' for European barrier options does not hold for American barrier options.
2. Recall the definition of the Macaulay duration in Section 1.6. How can the duration concept be applied to callable bonds? How does the value of a callable bond change if a flat yield curve is shifted up-/down-wards? What is the effect of permitting or forbidding early principal repayments (or: bond calls)?

Exercises with Mathematica and UnRisk

3. Assume a risk-free interest rate of 3 %. Determine the (constant) volatility of Apple's stock price so that the Apple knock-in reverse convertible has a fair price of 1,000 USD in the Black-Scholes model. To answer this question, use the UnRisk command `Make` to construct the embedded equity barrier option and use the command `Valuate` to price this option for a range of volatilities.
4. Plot the value of the down-and-in put option as a function of the stock price (between 20 and 45, at a point of valuation) with the volatility from Exercise 1. What happens as the valuation date approaches the maturity of the option? What is the Delta of the option? Reproduce Figure 13.2. What does the plot imply for possible hedging strategies?
5. Write the code of a Monte Carlo simulation to price a barrier option in the Heston model (with (discrete) daily barrier observations). Find a case where the Black-Scholes and the Heston price of a plain vanilla option are equal, while the respective prices of an up-and-out call option differ significantly.
6. Use the UnRisk commands `MakeFixedRateBond`, `MakeCallPutSchedule`, and `MakeCPFixedRateBond` to construct the callable bond of Section 13.2. Use `Properties` to verify that the input is correct. Construct a one-factor Hull-White model based on reasonable values of a flat interest curve, constant reversion speed and constant volatility. Price the callable bond.
7. Use UnRisk to construct the snowball floater of Section 13.3 (Libor is set in arrears). Run your computations for a non-callable bond, as well as for the case where the bond is callable on every coupon day starting with the third one. Assume that Libor on coupon setting dates can only take the values 2.5, 3.5 or 4.5 %. Determine the resulting coupon sizes for the non-callable snowball.
8. Assume a flat interest (zero) curve of 3.5 % (continuous compounding) and construct a Hull-White model with reversion speed 0.1 (per year) and an annual Hull-White volatility of 1 %. Determine the value of the non-callable snowball. Repeat your computations for the callable case. What is the effect on the value of (callable/non-callable) snowball, if the interest curve is shifted by +/− 50 or 100bps. How can this be interpreted as duration?

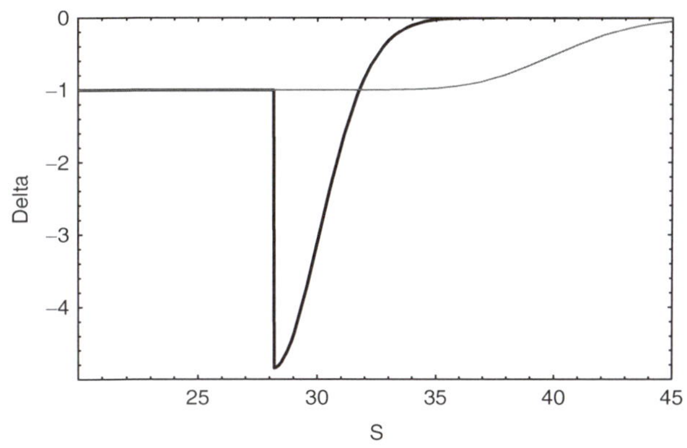

Fig. 13.2 Delta of the down-and-in-put option (black/bold) and the plain vanilla put option (blue, same strike), that is obtained once the barrier has been hit. Two weeks prior to maturity ($r = 0.03$, $\sigma = 0.35$)

9. Go back to Exercise 8 and increase the reversion speed to 1 or 10, respectively. How does this affect the value of the (callable/non-callable) snowball? Explain the source(s) of the change in value.
10. Determine the constant Hull-White volatility σ_1 such that the callable snowball with volatility σ_1 and reversion speed 1 has the same value as the non-callable snowball of Exercise 2. How would the callable values compare?
11. If the steepener on page 148 is not callable, what is the effect on its value of increasing the CMS spread by 25 basis points?
12. (Advanced). UnRisk also offers a tool to valuate contracts in a Libor market model (LMM). Try to obtain the required market data (interest curve, at-the-money cap volatilities, swaption volatilities) to calibrate an LMM and to valuate a steepener instrument. What is the difference in value if the steepener is callable or non-callable?
13. Produce a code for a Monte Carlo simulation (see Chapter 11) to generate sample paths of the prices of N stocks. Assume that the stock prices are uncorrelated and all have the same volatility σ. How does the number N of different stocks and the length of the coupon period drive the value of a 5-year lock-in swing?

Portfolio Optimization

14

Insurers, banks, mutual funds, sovereign wealth funds and also individuals invest money in the financial markets in order to generate financial returns. Hereby the investor *allocates* capital to different investments, such as low-risk low-expected return investments (e.g. high quality government bonds, bank accounts) or higher-risk higher-expected return investments (e.g. stocks, real estate, commodities). It is one of the core problems in finance to provide decision making tools for the optimal (or: *efficient*) allocation of capital. Optimality in this context depends on the decision maker's liquidity needs and risk aversion. This chapter will introduce the classical mean-variance optimization framework in a static one-period setup, and proceed to continuous-time portfolio optimization problems.

14.1 Mean-Variance Optimization

Mean-Variance Optimization, No Risk-Free Asset

Assume that an investor wishes to allocate funds w_0 (investable wealth) across n risky assets (e.g. stocks, bonds or real estate) that trade at the prices $\mathbf{P} = (P_1(t), P_2(t), \ldots, P_n(t))'$, $0 \le t \le T$, and that the current prices $\mathbf{P}(0) = \mathbf{p} = (p_1, p_2, \ldots, p_n)'$ are observable.[1] A portfolio $\mathbf{a}(t) = (a_1(t), a_2(t), \ldots, a_n(t))'$ holds a_i units of the i-th asset at time t, and portfolios here are assumed *static* over $[0, T]$ (e.g. a week, a month, a year), i.e. there is no re-allocation prior to T,

[1] In this chapter we will use a fair amount of vector/matrix notation to keep things compact. Vectors are printed in bold (e.g. **a**). $\mathbf{a}'$ is the transposed vector of **a** (e.g. if **a** has dimensions $n \times 1$ (a 'column' vector), then $\mathbf{a}'$ will be the corresponding $1 \times n$ matrix (a 'row') with $a_{i,1} = a'_{1,i}$ $(1 \le i \le n)$). For $n \times 1$ dimensional vectors **a** and **b**, we have $\mathbf{a}'\mathbf{b} = \sum_{i=1}^{n} a_i b_i$ (the so-called scalar product of **a** and **b**). We define $\mathbf{1} = (1, 1, \ldots, 1)'$ so that $\mathbf{a}'\mathbf{1} = \sum_i a_i$ is simply the sum of the elements of vector **a**. The inverse matrix of A (dimension $n \times n$) is denoted by A^{-1} and we have $A^{-1}A = AA^{-1} = I_n$, with I_n being the $n \times n$ identity matrix (i.e. '1's in the main diagonal, and '0' entries otherwise).

H. Albrecher et al., *Introduction to Quantitative Methods for Financial Markets*, Compact Textbooks in Mathematics, DOI 10.1007/978-3-0348-0519-3_14,

so $a_i(t) = a_i(0)$ for $0 \leq t \leq T$. To shorten notation, we will simply write $\mathbf{a}$ in the following. The portfolio is required to fulfil the *initial budget constraint*

$$\sum_{i=1}^{n} a_i p_i = \mathbf{a}'\mathbf{p} = w_0,$$

and the investor's (random) wealth at time T is

$$W(T) = \sum_{i=1}^{n} a_i P_i(T) = \mathbf{a}'\mathbf{P}(T).$$

Different allocations $\mathbf{a}$ will now give different final wealth distributions $W(T)$, and one will look for a decision criterion of how to choose $\mathbf{a}$. Note that $\mathbf{a}$ depends on w_0 and on the prices p_i, so that in the following we prefer to state a portfolio in terms of *percentage weights* $\boldsymbol{\theta} = (\theta_1, \theta_2, ..., \theta_n)'$ of the initial wealth w_0, with $\theta_i = \frac{a_i \cdot p_i}{w_0}$. Also, we will translate the price change from $\mathbf{p}$ to $\mathbf{P(T)}$ into returns $\mathbf{R} = (R_1, R_2, ..., R_n)'$, with $P_i(T) = (1 + R_i) \cdot p_i$, so that the final wealth can also be written as

$$W(T) = w_0 \cdot \sum_{i=1}^{n} \theta_i (1 + R_i) = w_0 \cdot \boldsymbol{\theta}'(\mathbf{1} + \mathbf{R}),$$

with $R_{\text{pf}} = \sum_{i=1}^{n} \theta_i R_i = \boldsymbol{\theta}'\mathbf{R}$ being the *return of the* entire investment *portfolio*.

Classical Portfolio Theory as introduced by Markowitz[2] suggests that investors trade off expected return against risk described through the return variance, and the portfolio allocation problem (Markowitz problem (MP)) is written as[3]

$$\begin{aligned} (\text{MP1}):\quad & \max_{\boldsymbol{\theta}}\big(\mathbb{E}[R_{\text{pf}}] - \lambda \cdot \text{Var}[R_{\text{pf}}]\big) \\ & \big(\max_{\boldsymbol{\theta}} \mathbb{E}[\boldsymbol{\theta}'\mathbf{R}] - \lambda \cdot \text{Var}[\boldsymbol{\theta}'\mathbf{R}]\big) \qquad (14.1) \\ & \text{subject to} \quad \boldsymbol{\theta}'\mathbf{1} = 1,\ \lambda \geq 0. \end{aligned}$$

The choice of the parameter λ will depend on the risk aversion of the investor, and a higher risk aversion will result in a larger λ (that penalizes more for portfolio return variance). The constraint on $\boldsymbol{\theta}$ ensures that exactly the initial wealth w_0 is invested at the beginning. The Markowitz problem (14.1) has two equivalent formulations,

$$(\text{MP2}):\quad \min \text{Var}[\boldsymbol{\theta}'\mathbf{R}] \quad \text{subject to } \mathbb{E}[\boldsymbol{\theta}'\mathbf{R}] = \bar{r},\ \boldsymbol{\theta}'\mathbf{1} = 1$$

---

[2]Harry Markowitz (1927–) was awarded the 1990 Nobel Memorial Prize in Economic Sciences, together with Merton H. Miller (1923–2000) and William F. Sharpe (1934–), 'for his pioneering work in the theory of financial economics'.

[3]This formulation is equivalent to $\max_{\mathbf{a}}(\mathbb{E}[W(T)] - \text{Var}[W(T)])$. In the sequel, transaction costs and taxes are neglected.

and

$$\text{(MP3)}: \quad \max \mathbb{E}[\boldsymbol{\theta}'\mathbf{R}] \quad \text{subject to } \operatorname{Var}[\boldsymbol{\theta}'\mathbf{R}] = \bar{\sigma}^2,\ \boldsymbol{\theta}'\mathbf{1} = 1.$$

In (MP2), the portfolio with the lowest return variance is chosen among all portfolios of expected return $\bar{r}$. Conversely, solving (MP3) gives the maximum mean-return portfolio at a certain variance level $\bar{\sigma}^2$. When deciding based on (MP2) or (MP3), the investor's aversion to risk again plays a role, as it will drive the desired $\bar{r}$ or $\bar{\sigma}^2$ level (and each $\bar{r}$ (or: $\bar{\sigma}^2$) implies a particular $\lambda = \bar{\lambda}$ in (MP1)). The equivalence of (MP1)–(MP3) in terms of returning the same optimal mean-variance combinations becomes clear from using the method of Lagrange multipliers to restate the three optimization problems.

Denote the mean return vector of the assets as $\boldsymbol{\mu} = (\mu_1, \mu_2, ..., \mu_n)'$ (with $\mu_i := \mathbb{E}[R_i]$). The mean return of a portfolio $\boldsymbol{\theta}$ is then

$$\mu_{\text{pf}} = \mathbb{E}[R_{\text{pf}}] = \sum_{i=1}^{n} \theta_i \mathbb{E}[R_i] = \boldsymbol{\theta}'\boldsymbol{\mu}$$

and the variance of the portfolio return is

$$\sigma_{\text{pf}}^2 = \operatorname{Var}[R_{\text{pf}}] = \sum_{i=1}^{n}\sum_{j=1}^{n} \theta_i\theta_j \operatorname{Cov}[R_i, R_j] = \boldsymbol{\theta}'\,\boldsymbol{\Sigma}\,\boldsymbol{\theta},$$

with the covariance matrix $\Sigma = (\Sigma_{ij})_{1 \le i,j \le n}$, i.e.

$$\Sigma_{ij} = \operatorname{Cov}[R_i, R_j] = \mathbb{E}\big[(R_i - \mu_i)\,(R_j - \mu_j)\big].$$

Note in this context that Markowitz optimization only requires the knowledge of the first two moments of the asset return vector **R**, but not the specification of its distribution.[4] To solve (MP1), we can use the method of Lagrange multipliers and set the partial derivatives of the Lagrange function

$$L(\boldsymbol{\theta}, \gamma) = \boldsymbol{\theta}'\boldsymbol{\mu} - \lambda \cdot \boldsymbol{\theta}'\Sigma\boldsymbol{\theta} - \gamma \cdot (\boldsymbol{\theta}'\mathbf{1} - 1)$$

with respect to $\boldsymbol{\theta}$ and γ equal zero.[5] This gives the necessary (and here sufficient) conditions on the solution $(\boldsymbol{\theta}^*, \gamma^*)$,

[4]The mean and the variance parameters capture all features of the return distribution in the multivariate normal case, however, this will not necessarily be the case for other distribution classes (e.g. which are skewed or have relatively more tail mass).

[5]Whoever is not used to matrix algebra can also understand the below formulas componentwise, e.g. the first vector component of $\partial L/\partial\boldsymbol{\theta}$ is simply $\partial L/\partial\theta_1$, etc. Obtaining the solutions to the mean-variance problem in matrix form will greatly facilitate the implementation in computer programs for larger n.

$$\left.\frac{\partial L}{\partial \boldsymbol{\theta}}\right|_{\boldsymbol{\theta}=\boldsymbol{\theta}^*,\gamma=\gamma^*} = \boldsymbol{\mu} - 2\lambda \cdot \Sigma\boldsymbol{\theta}^* - \gamma^* \cdot \mathbf{1} = 0$$

and

$$\left.\frac{\partial L}{\partial \gamma}\right|_{\boldsymbol{\theta}=\boldsymbol{\theta}^*,\gamma=\gamma^*} = \boldsymbol{\theta}^{*\prime}\mathbf{1} - 1 = 0.$$

Separating $\boldsymbol{\theta}^*$ in the first equation and substituting this expression into the second condition gives $\gamma^* = \frac{\mathbf{1}'\Sigma^{-1}\boldsymbol{\mu}-2\lambda}{\mathbf{1}'\Sigma^{-1}\mathbf{1}}$, so that one ultimately finds (by substituting this back into the first condition)

$$\boldsymbol{\theta}^* = \frac{\Sigma^{-1}\mathbf{1}}{\mathbf{1}'\Sigma^{-1}\mathbf{1}} + \frac{1}{2\lambda}\Sigma^{-1}\left(\boldsymbol{\mu} - \mathbf{1}\frac{\mathbf{1}'\Sigma^{-1}\boldsymbol{\mu}}{\mathbf{1}'\Sigma^{-1}\mathbf{1}}\right). \tag{14.2}$$

This formula is also known as *Two Fund Separation Theorem,* since the weights $\boldsymbol{\theta}^*$ are stated as a sum of two terms. The first term depends neither on $\boldsymbol{\mu}$ nor on λ (so that this term is the same for every investor), while the weight adjustment from the second term is driven by the investor's choice of λ (in line with his risk willingness). If an investor only accepts as little return variance as possible (i.e. $\lambda \to \infty$), we find that

$$\boldsymbol{\theta}^*_{\text{mv}} = \frac{\Sigma^{-1}\mathbf{1}}{\mathbf{1}'\Sigma^{-1}\mathbf{1}},$$

and one is dealing with what is also referred to as *minimum variance portfolio.*

Example (Markowitz portfolio)

Consider an investment universe of three assets with mean returns $\boldsymbol{\mu} = (7.9\%, 9.0\%, 7.1\%)'$, standard deviations $\boldsymbol{\sigma} = (18.2\%, 18.3\%, 16.5\%)'$ and correlation coefficients $\rho_{12} = 0.47$, $\rho_{13} = 0.14$ and $\rho_{23} = 0.25$.[6] We compute

$$\Sigma^{-1} = \begin{bmatrix} 38.78 & -17.89 & -1.03 \\ -17.89 & 40.11 & -8.36 \\ -1.03 & -8.36 & 39.21 \end{bmatrix}$$

and then (14.2) gives the vector of optimal portfolio weights

$$\boldsymbol{\theta}^* = \begin{bmatrix} 0.3125 \\ 0.2181 \\ 0.4694 \end{bmatrix} + \frac{1}{2\lambda}\begin{bmatrix} -0.1617 \\ 0.5268 \\ -0.3652 \end{bmatrix}$$

as a function of the risk-aversion parameter λ. Furthermore $\mathbb{E}[R_{\text{pf}}] = \boldsymbol{\mu}'\boldsymbol{\theta}^* = 0.0776 + \frac{1}{\lambda}0.0044$.

[6] The covariance matrix is calculated in this case as $\Sigma = \boldsymbol{\sigma}'\mathbf{I}_3(\rho_{ij})\mathbf{I}_3\sigma$, where $\mathbf{I}_3$ is the 3×3 identity matrix and (ρ_{ij}) is the correlation matrix.

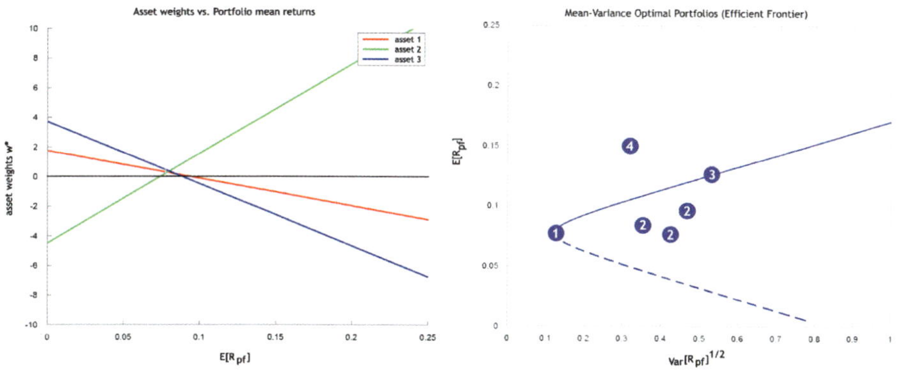

Fig. 14.1 Portfolio weights $\boldsymbol{\theta}^*$ as function of $\mathbb{E}[R_{\text{pf}}]$ (left), efficient frontier in the mean-standard deviation plane (right)

The minimum variance portfolio is given by $\boldsymbol{\theta}^*_{\text{mv}} = (0.3125, 0.2181, 0.4694)'$ with mean return $\mathbb{E}[R_{\text{pf}}]_{\text{mv}} = 0.0776$. As λ decreases (i.e. an investor is willing to accept more return variance), the relative allocation to the second asset (which shows the highest mean return) increases due to its positive weight of 0.5268 in the λ-term of $\boldsymbol{\theta}^*$, and the expected return grows to compensate for the additional variance.

The left graph in Figure 14.1 shows the θ_i^*'s as linear functions of the expected portfolio return. The solid line in the figure on the right is called *efficient frontier* and indicates optimal $(\mu_{\text{pf}}, \sigma_{\text{pf}})$ - combinations which can be attained by varying λ. The dashed line is obtained by solving (MP1), but with $\lambda < 0$. (1) is the minimum-variance portfolio, (2) are *admissible* (mostly non-optimal) portfolios that can be attained for some $\boldsymbol{\theta}$ and are bounded by the solid and the dashed line, (3) is a mean-variance optimal portfolio and (4) cannot be attained based on the three given assets.

Mean-Variance Optimization, With a Risk-free Asset

Suppose now a risk-free asset that returns r_{f} (with $r_{\text{f}} < \min_i \mu_i$) over $[0, T]$ is added to the set of possible investments.[7] The difference $w_0 \cdot (1 - \sum_{i=1}^n \theta_i)$ is invested in (if positive) or borrowed from (if negative) the risk-free account, so that the portfolio return is now $R_{\text{pf}} = \boldsymbol{\theta}'\mathbf{R} + r_{\text{f}}(1 - \boldsymbol{\theta}'\mathbf{1})$ and the mean-variance optimization problem (14.2) becomes

$$\max_{\boldsymbol{\theta}} \big(\boldsymbol{\theta}'\boldsymbol{\mu} + r_f(1 - \boldsymbol{\theta}'\mathbf{1}) - \lambda \cdot \boldsymbol{\theta}'\Sigma\boldsymbol{\theta}\big). \tag{14.3}$$

Note that no budget restriction is needed here, so that $\lambda \geq 0$ is the only optimization constraint. Setting the first derivative of the objective function with respect to $\boldsymbol{\theta}$ equal to zero returns the optimal portfolio weights as

[7] This extension of the initial Markowitz formulation of the problem for only risky assets was suggested by James Tobin (1918–2002), who was awarded the Nobel Memorial Prize in Economic Sciences in 1981 for his contributions in the field of portfolio theory.

$$\boldsymbol{\theta}^* = \frac{1}{2\lambda}\Sigma^{-1}(\boldsymbol{\mu} - r_f \cdot \mathbf{1}). \tag{14.4}$$

It is then straightforward to obtain

$$\mu_{\rm pf}^* = \mathbb{E}[R_{\rm pf}^*] = \boldsymbol{\theta}^{*\prime}\boldsymbol{\mu} + r_{\rm f}(1 - \boldsymbol{\theta}^{*\prime}\mathbf{1}) = r_{\rm f} + \frac{1}{2\lambda}(\boldsymbol{\mu} - r_{\rm f}\mathbf{1})'\Sigma^{-1}(\boldsymbol{\mu} - r_{\rm f}\mathbf{1})$$

and

$$\sigma_{\rm pf}^{*2} = {\rm Var}[R_{\rm pf}^*] = \boldsymbol{\theta}^{*\prime}\Sigma\boldsymbol{\theta}^* = \left(\frac{1}{2\lambda}\right)^2 (\boldsymbol{\mu} - r_{\rm f}\mathbf{1})'\Sigma^{-1}(\boldsymbol{\mu} - r_{\rm f}\mathbf{1}).$$

Solving the last equation for $\frac{1}{2\lambda}$ and substituting the result in the optimal mean return finally gives

$$\mu_{\rm pf}^* = r_{\rm f} + \sqrt{(\boldsymbol{\mu} - r_{\rm f}\mathbf{1})'\Sigma^{-1}(\boldsymbol{\mu} - r_{\rm f}\mathbf{1})} \cdot \sigma_{\rm pf}^*,$$

so that the mean-variance optimal $(\mu_{\rm pf}^*, \sigma_{\rm pf}^*)$-combinations are now described by a straight line, the so-called *capital markets line* (CML). In the $(\mu_{\rm pf}, \sigma_{\rm pf})$-plane, the CML runs through the risk-free investment point $(r_{\rm f}, 0)$ and touches the efficient frontier (as derived for the original problem with only risky assets) in exactly one point, the so-called *market portfolio* (or: *tangency portfolio*) $\boldsymbol{\phi} = (\phi_1, \phi_2, ..., \phi_n)$. Since for the market portfolio all wealth is invested in risky assets, conditioning on $\boldsymbol{\theta}'\mathbf{1} = 1$ in (14.3) yields the market portfolio weights

$$\boldsymbol{\phi} = \frac{\Sigma^{-1}(\boldsymbol{\mu} - r_{\rm f}\mathbf{1})}{\mathbf{1}'\Sigma^{-1}(\boldsymbol{\mu} - r_{\rm f}\mathbf{1})},$$

with mean return and return standard deviation

$$\mu_\phi = \boldsymbol{\mu}'\boldsymbol{\phi} = \frac{\boldsymbol{\mu}'\Sigma^{-1}(\boldsymbol{\mu} - r_{\rm f}\mathbf{1})}{\mathbf{1}'\Sigma^{-1}(\boldsymbol{\mu} - r_{\rm f}\mathbf{1})} \text{ and } \sigma_\phi = \sqrt{\boldsymbol{\phi}'\Sigma\boldsymbol{\phi}} = \frac{\sqrt{(\boldsymbol{\mu} - r_{\rm f}\mathbf{1})'\Sigma^{-1}(\boldsymbol{\mu} - r_{\rm f}\mathbf{1})}}{\mathbf{1}'\Sigma^{-1}(\boldsymbol{\mu} - r_{\rm f}\mathbf{1})}.$$

With a risk-free asset, every mean-variance optimizing investor chooses the optimal investment that is simply a (linear) combination of the risk-free investment and the market portfolio. The more risk averse the investor is (i.e. for higher λ), the more she will allocate to the risk free asset (cf. (14.4)). This will now be further illustrated in the following example.

Example (Mean-variance portfolio selection, with a risk-free asset)

Use the risky assets from the Markowitz portfolio example in the previous section, and add a risk-free asset that returns $r_f = 1\%$. We compute the weights of the market portfolio as $\boldsymbol{\phi} = \frac{\Sigma^{-1}(\boldsymbol{\mu} - r_{\rm f}\mathbf{1})}{\mathbf{1}'\Sigma^{-1}(\boldsymbol{\mu} - r_{\rm f}\mathbf{1})} = (0.275, 0.341, 0.384)'$. Simply investing in the market portfolio would give the expected return

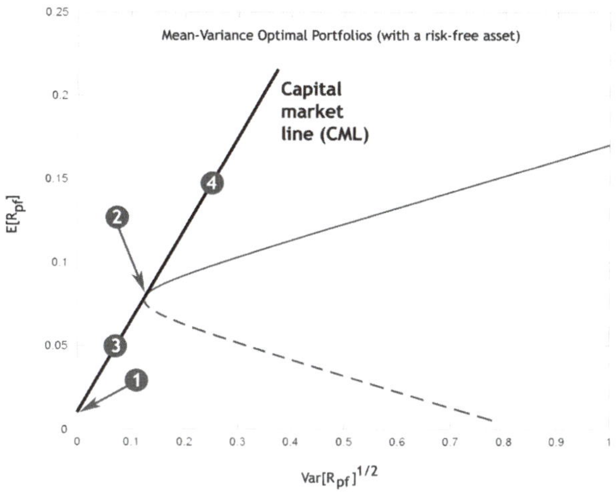

Fig. 14.2 Risky and risk-free assets: capital market line and efficient frontier

$$\mu_{\boldsymbol{\phi}} = \boldsymbol{\mu}'\boldsymbol{\phi} = (0.079, 0.09, 0.071)'(0.275, 0.341, 0.384) = 7.97\%.$$

Assume an investor aims to have a mean return of $\mathbb{E}[R_{\text{pf}}] = 5\%$, in line with his risk preferences. Since $r_f = 1\% < 5\% < 7.97\% = \mu_{\boldsymbol{\phi}}$, the optimal portfolio will lie on the capital market line somewhere between the risk-free investment and the market portfolio. Solving

$$5\% = (1-\alpha)1\% + \alpha 7.97\%$$

yields $\alpha = 0.5739$. The investor will hence invest $0.4261 w_0$ in the risk-free asset, and $w_0\boldsymbol{\theta}^*$ in the three risky assets with $\boldsymbol{\theta}^* = 0.5739 \cdot (0.275, 0.341, 0.384)' = (0.158, 0.196, 0.221)'$. The standard deviation of this portfolio is then given by $\sigma_{\text{pf}} = \sqrt{\boldsymbol{\theta}^{*\prime}\Sigma\boldsymbol{\theta}^*} = 0.073$. Alternatively, one could have the same portfolio by specifying $\lambda = 3.745$ in (14.4).

In Figure 14.2, (1) is the risk-free portfolio, (2) is the market (or: tangency) portfolio, (3) is the portfolio derived in this example with $\mathbb{E}[R_{\text{pf}}] = 5\%$ and (4) is a *levered* portfolio with $\boldsymbol{\theta}'\mathbf{1} > 1$ so that $\boldsymbol{\theta}'\mathbf{1} - 1$ is borrowed from the risk-free account to finance the purchase of risky assets in excess of w_0. From the plot it becomes also clear that the capital market line can be written as

$$\mu^*_{\text{pf}} = r_{\text{f}} + \frac{\mu_{\boldsymbol{\phi}} - r_{\text{f}}}{\sigma_{\boldsymbol{\phi}}} \cdot \sigma^*_{\text{pf}},$$

and the market portfolio is found to be the efficient portfolio with the greatest *Sharpe ratio* $\frac{\mu - r_{\text{f}}}{\sigma}$.[8]

Finally note that the covariance between the i-th risky asset and the market portfolio is

$$\text{Cov}(R_i, R_{\boldsymbol{\phi}}) = (0, \ldots, 0, 1, 0, \ldots, 0)'\Sigma\boldsymbol{\phi} = \frac{\mu_i - r_{\text{f}}}{\mathbf{1}'\Sigma^{-1}(\boldsymbol{\mu} - r_{\text{f}}\mathbf{1})} = \frac{(\mu_i - r_{\text{f}})\sigma^2_{\boldsymbol{\phi}}}{\mu_{\boldsymbol{\phi}} - r_{\text{f}}},$$

[8]The Sharpe ratio is a popular measure to compare investments, as it divides the mean return in excess of the risk-free rate ($\mu - r_{\text{f}}$, also: risk premium) by a number related to the involved risk (σ).

which immediately leads to

$$\mathbb{E}[R_i] = \mu_i = r_f + \beta_i(\mu_\phi - r_f), \tag{14.5}$$

where the *beta coefficient* is defined as $\beta_i := \mathrm{Cov}(R_i, R_\phi)/\sigma_\phi^2$.

Equation (14.5) is the core relation used in the *Capital Asset Pricing Model* (short: CAPM).[9] CAPM considers one market with multiple investors, with all investors selecting their portfolios based on mean-variance optimization. Asset prices $p_1, \ldots, p_n$ in this model are based on mean return and return covariance parameters in a demand-supply equilibrium (which is obtained if the composition of the market is identical to the market portfolio ϕ).

Remark 14.1. Equation (14.5) can be understood as a *linear regression* of the mean asset returns against the mean return of the market portfolio as explanatory variable. β_i is the regression coefficient and explains the sensitivity of the i-th asset to the entire market. The risk aggregated in the market portfolio is often referred to as *systematic risk* and cannot be further reduced through diversification.

Additional Constraints, Transaction Costs, Parameter Estimation

Weight Constraints and Transaction Costs

Let us now return to mean-variance optimization with only risky assets. An investment fund, for example, could have the task of selecting its portfolio from a certain set of stocks. The investors in the fund could then decide themselves how much to hold in risky fund units and how much to allocate to a risk-free account. Note that $\boldsymbol{\theta} \in \mathbb{R}$ without any restrictions up to this stage, and transaction costs were also neglected. There are several constraints that might be relevant to an investor. For example, a *long-only* constraint states that assets can only be bought but not short-sold, i.e. $\theta_j \geq 0$ for all $1 \leq i \leq n$. *Holding constraints* limit the portfolio exposure to specific assets (e.g. $\theta_i \leq u_i$) or subsets of stocks (e.g. $I_j \subset \{1, 2, \ldots, n\}$ is a subset of the n stocks, so that $\sum_{i \in I_j} \theta_i \leq u_j$). *Cardinality constraints* restrict the maximum number x of different assets in a portfolio, which have to be monitored by the investor (e.g. define $\delta_i = \mathbf{1}_{\{\theta_i \neq 0\}}$ and set the constraint $\sum_{i=1}^n \delta_i \leq x$).

While *transaction costs* have been neglected up to this point, it will be the typical situation that some initial portfolio $\boldsymbol{\theta}_{\text{in}}$ has to be adjusted (or: re-optimized) and the buying and selling of stocks will incur transaction costs. Suppose $\boldsymbol{\theta}^+$ is the positive change to the weights due to purchases, while $\boldsymbol{\theta}^-$ reflects the asset sales so that the portfolio weights $\boldsymbol{\theta}$ after the adjustment are

$$(\text{C1}): \quad \boldsymbol{\theta} = \boldsymbol{\theta}_{\text{in}} + \boldsymbol{\theta}^+ - \boldsymbol{\theta}^-.$$

[9] The CAPM was initially developed by William Sharpe (1964), see the footnote on page 156, and John V. Lintner (1965).

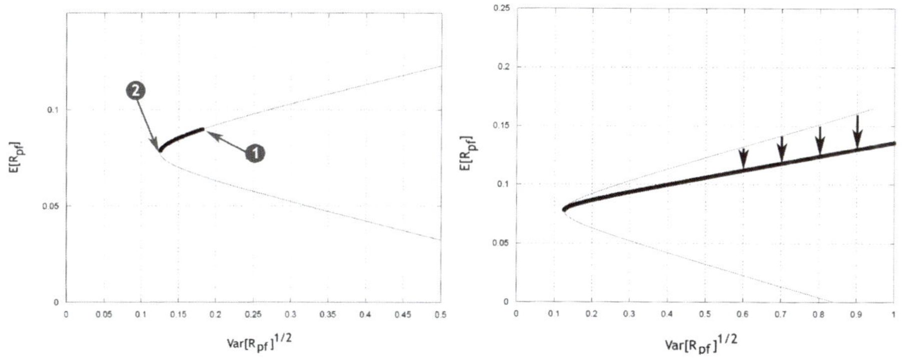

Fig. 14.3 Long-only portfolio (left) and effect of transaction costs with the minimum-variance portfolio as initial portfolio (right)

Proportional cost factors $c_i^+, c_i^- \geq 0$ describe the transaction costs from buying or selling assets as $c_i^+\theta_i^+$ and $c_i^-\theta_i^-$ so that the reduction in wealth from transaction costs results in $\sum_{i=1}^n \theta_i \leq 1$; in particular

$$(\mathrm{C2}): \quad \boldsymbol{\theta}'\mathbf{1} = 1 - \boldsymbol{\theta}^{+\prime}\mathbf{c}^+ - \boldsymbol{\theta}^{-\prime}\mathbf{c}^- .$$

Finally, the objective function differs from (14.1) as the transaction costs reduce the mean return,

$$\max_{\boldsymbol{\theta}^+,\boldsymbol{\theta}^-} \left\{ (\boldsymbol{\theta}'(\mathbf{1}+\boldsymbol{\mu}) - 1) - \lambda \cdot \boldsymbol{\theta}'\Sigma\boldsymbol{\theta} \right\}.$$

Although the above optimization problem under transaction costs with the conditions (C1), (C2) and $\boldsymbol{\theta}^+, \boldsymbol{\theta}^- \geq 0$ looks more complex, it is still of quadratic form with linear constraints and finding a solution by numerical optimization algorithms (e.g. in Mathematica) is straightforward.

The left graph in Figure 14.3 shows the effect of restricting the portfolio weights by $\boldsymbol{\theta} \geq 0$, using the assets in the above example.[10] The (restricted) efficient frontier now runs only from the minimum-variance portfolio (2) to the portfolio (1) of only the asset with the highest mean return (i.e. with $\theta_i = 1$ for $\mu_i = \max_{1 \leq j \leq n} \mu_j$ and $\theta_j = 0$ otherwise). The plot on the right shows the effect of transaction costs if one initially holds the minimum-variance portfolio. For a given standard deviation, the mean return under transaction costs is now lower. This effect becomes stronger the further one moves away from the initial portfolio.

[10]In general, the efficient frontier with the long-only constraint will run *below* the unconstrained efficient frontier.

Parameter Estimation and Comments

In practice, it is often difficult to attain suitable parameter estimates $\widehat{\mu}$ and $\widehat{\Sigma}$. Empirical studies (cf. Chopra & Ziemba [16]) suggest that the mean-variance framework is particularly sensitive to small changes in the mean estimates, as small changes in the applied model parameters can lead to large changes in the optimal weights $\widehat{\boldsymbol{\theta}}^*$. Estimation issues are addressed by many publications dealing with robust portfolio selection (e.g. Fabozzi et al. [30] or Scherer & Martin [68]), and a much cited article by Black & Litterman [8] suggests a procedure to blend investor views and historical data for the parameter estimation to mitigate the issue that generally historical parameter estimates do not reflect investor views.

Another shortcoming of the mean-variance framework is that the variance is often seen as an inadequate risk measure to capture the dependence structure between asset returns and hence the features of the portfolio return distribution.[11] Furthermore, a static one-period model does not allow for rebalancing of the portfolio prior to T. All these drawbacks have motivated researchers to develop alternative strategies for portfolio selection. Nevertheless, the Markowitz framework is a milestone in the development of Portfolio Theory and mean-variance optimal portfolios are often derived in practice as a first benchmark or for comparative purposes.

14.2 Risk Measures and Utility Theory

In mean-variance optimization, risk is only measured by the variance or the standard deviation, and the 2007/08 credit crisis is only one of the events that fueled discussions of what risk measures are best suited in practice. Measuring risk is of great importance for financial firms (e.g. banks and insurers), which have to comply with regulatory (capital) requirements based on their risk portfolio to ensure prudent conduct of business and a stable economy. In the EU, capital requirements for banks are regulated in the Basel rules (Basel I, II and III) and insurance companies have to comply with Solvency regulation (the Solvency I and II Directives).

Let us define a *risk* X as a non-negative random variable, i.e. $\mathbb{P}[X \geq 0] = 1$, which describes some random loss. The question of how to measure risk is now closely related to ordering different risks by preference. For instance, for two risks X_1 and X_2, an individual would typically be able to give a preference ('better') or to state indifference ('equally good'). For such comparison, one can reduce the features of a risk distribution to one number and call it *risk measure*:

Definition (Risk measure). A *risk measure* ρ is a mapping that assigns a particular monetary value to each risk, i.e. $\rho(X) = x$ EUR, $x \in \mathbb{R}$.

[11]Positive deviations (excess profits) from the mean return are equally penalized as negative deviations. In connection to this, the idea of using a risk measure that only punishes for negative deviations from the mean return already goes back to Markowitz.

Economically the risk measure could be linked to the capital one would need to hold as buffer against adverse outcomes of the risk. The standard deviation (or: variance) is a well-known, yet often unsatisfactory risk measure. Another risk measure is the *value-at-risk*:

Definition. The *value-at-risk* $\mathrm{VaR}(X)$ of a risk X at some confidence level α is defined as the α-quantile of the distribution of X, i.e.

$$\mathrm{VaR}_\alpha(X) := \inf\{x : \mathbb{P}[X \leq x] \geq \alpha\}. \tag{14.6}$$

For example, $\mathrm{VaR}_{0.99}(X) = x$ of an investment implies that x will be sufficient to compensate the loss X in 99% of the cases. $\mathrm{VaR}_\alpha(X)$ is a popular risk measure as it is intuitive and relatively easy to compute.[12] However, the shortcomings of the value-at-risk include that it does not provide information on the conditional distibution of the loss X given that $X > \mathrm{VaR}_\alpha(X)$ (i.e. on the tail of the distribution). Moreover, value-at-risk is not *sub-additive*. Sub-additivity of a risk measure ρ means that for any two risks X_1 and X_2, it holds that

$$\rho(X_1 + X_2) \leq \rho(X_1) + \rho(X_2)$$

(see Exercise 4). Sub-additivity hence reflects the intuition that the risk of a portfolio is not greater than the sum of the risks of its parts (*diversification of risk*).

The standard deviation is an example of a sub-additive risk measure, and so is the *expected shortfall* (ES). The latter is related to the value-at-risk and is defined for some confidence level α as

$$\mathrm{ES}_\alpha(X) = \mathbb{E}[X|X > \mathrm{VaR}_\alpha] = \frac{1}{1-\alpha}\int_{\mathrm{VaR}_\alpha}^{\infty} x\, dF_X(x),$$

where $F_X(x)$ is the c.d.f. of the risk X.

The expected shortfall ES_α is the expectation of the loss, given that it exceeds the level VaR_α. Based on only few empirical data points on extreme losses, the estimation of ES_α often raises more issues than that of the VaR. However, given its favorable mathematical properties, the expected shortfall is often used in practice, for example, in the Swiss Solvency Test (SST) for insurance companies. Finally, note that ES_α is a so-called *coherent risk measure* (see Exercise 4).

An alternative way of expressing preferences between risks is given by *utility theory*. In utility theory, an investor can describe his utility $u(x)$ from certain levels of wealth x. The motivation for this rich approach was provided by a result of John von Neumann and Oskar Morgenstern[13] which reads as follows (under some

[12]For example, both the Basel II and Solvency II accords define value-at-risk as a standard risk measure.

[13]John von Neumann (1903–1957) and Oskar Morgenstern (1902–1977) are also often credited for having built the foundation of *game theory*.

fairly general consistency assumptions): when an investor can express a preference or indifference regarding any two investments X and Y, there exists a *utility function* u, such that $\mathbb{E}[u(X)] > \mathbb{E}[u(Y)]$ if and only if X is preferred over Y, and vice versa. In this way, the problem of finding the most-preferred investment strategy reduces to finding the strategy that maximizes the expected utility of the investor's wealth (an example is given in Section 14.3).

The determination of an adequate utility function is essential, yet challenging. In practice one can only approximate such a utility function. An investor is called *risk-averse* if the used utility function is (a) monotone increasing (more money is preferred over less) and (b) concave (additional units of wealth bring relatively less additional utility as the investor becomes richer). Popular choices of utility functions are logarithmic ($u(x) = \log(x)$), exponential ($u(x) = 1 - \mathrm{e}^{-ax}$ with $a > 0$) and partial-power ($u(x) = x^{1-a}/(1-a)$ with $a > 0, a \neq 1$) utility functions. As an alternative to mean-variance portfolio selection, one could now solve $\max_\theta \mathbb{E}[u(W(T))]$ for a suitable utility function $u(x)$.

14.3 Portfolio Optimization in Continuous Time

Section 14.1 discussed portfolio optimization in a one-period setup. It is now natural to search for rules of optimal allocation in a multi-period or, more generally, in a continuous-time setting (i.e. the weights can be dynamically re-balanced, similar to Δ-hedging in the Black-Scholes model). This problem is significantly more complex than its one-period counterpart. However, for an investment universe of only one risky and one risk-free asset, the solution for an investor with a logarithmic utility function is surprisingly simple and was found by R. Merton in 1969 (this problem is hence known as the *Merton problem*). Despite the solution of this problem requiring techniques from *Optimal Control* which are beyond the scope of this text, we will outline the idea of a heuristic solution here.

The price S_t of the risky asset is now modeled by a geometric Brownian motion, while the risk-free asset B_t earns interest at some constant rate r, i.e.

$$dS_t = S_t\left(\mu dt + \sigma dW_t\right), \qquad dB_t = B_t\, r dt.$$

Let w_t be the fraction of the capital X_t at time t that is invested in the stock. The local dynamics of X_t is then given by

$$dX_t = X_t((1 - w_t)dB_t + w_t dS_t).$$

In the case of logarithmic utility $U_t = \log(X_t)$, Itô's Lemma leads to

$$U_T = \log(X_0) + \int_0^T ((1 - w_t)r + w_t\mu - w_t^2\sigma^2/2)dt + \int_0^T w_t\sigma\, dW_t.$$

The investor will now maximize her expected utility of the capital (here: of its increase)

$$\mathbb{E}[U_T] = \log(X_0) + \int_0^T \big(r + w_t(\mu - r) - w_t^2\sigma^2/2\big)dt + \mathbb{E}\left[\int_0^T w_t\sigma\, dW_t\right] \tag{14.7}$$

$$= \log(X_0) + \int_0^T \big(r + w_t(\mu - r) - w_t^2\sigma^2/2\big)dt.$$

To see the second equality in (14.7), note that for every fixed discretization of the second (stochastic) integral (cf. Chapter 6), the expected value in every discretization interval is zero, and the interchange of the integral and the expectation can be justified. The weight w_t is then independent of ΔW_t, which is intuitive since one cannot anticipate how W_t will develop. The optimal investment strategy w_t^* can eventually be determined by maximizing the quadratic function

$$\left(r + w_t^*(\mu - r) - w_t^{*2}\sigma^2/2\right) = \max_{w_t}\big(r + w_t(\mu - r) - w_t^2\sigma^2/2\big)$$

pointwise for every t. Since $\sigma^2 > 0$, w_t^* is optimal if the first derivative becomes zero, and we find

$$w_t^* = \frac{\mu - r}{\sigma^2},$$

which relates the risk premium $\mu - r$ of an investment to its risk σ^2.

We can see that the optimal strategy is intuitive, in particular it is given by keeping the proportion of the capital invested in the risky investment constant over the entire investment period. Although this strategy seems easy to implement, it requires the ability to trade continuously.

14.4 Key Takeaways, References and Exercises

Key Takeaways

After working through this chapter you should understand and be able to explain the following terms and concepts:

- Mean-variance optimization (only risky assets), three equivalent formulations of the Markowitz problem, Two-Fund Separation theorem, efficient frontier, impact of choice of λ (e.g. $\lambda \to 0, \infty$), minimum-variance portfolio, admissible portfolio
- Mean-variance optimization (with risk-free asset), mean-variance problem (no budget constraint), capital market line, optimal portfolios as linear combination of the risk-free asset and the market portfolio, Sharpe ratio, CAPM

- Long-only, holding and cardinality constraints
- Mean-variance problem under proportional transaction costs
- Risk measures, value-at-risk, sub-additivity, expected shortfall
- Expected utility maximization: utility functions, risk preferences
- Continuous time portfolio optimization: Merton problem

References

Due to the concise form of the book, this chapter has only introduced some of the most classical ideas in portfolio optimization to show the flavor of the topic. Since its beginnings, this field of research has developed at fast speed, and detailed discussions of the wide scope of results are given, for example, by Pflug & Römisch [62], Fernholz [32], Platen & Heath [63], Dana & Jeanblanc [19], Korn & Korn [47], Luenberger [55], or Karatzas & Shreve [46].

Exercises

1. Show that the efficient frontier in the Markowitz model is convex. (Hint: apply the Cauchy-Schwartz inequality to some linear combination of two portfolios.)
2. In the Black-Scholes model, one finds for small Δt that

$$\Delta S_t = (S_{t+\Delta t} - S_t) \approx S_t(\mu \Delta t + \sigma \Delta W_t).$$

   Use this approximation to show that the capital market line in the Black-Scholes model (where we consider only one stock and the risk-free asset) has the slope $\sigma/(\mu - r)$ for the period $(t, t + \Delta t)$. Explain why the Sharpe ratio $(\mu - r)/\sigma$ is also called market price of risk in the Black-Scholes model.
3. Show that (14.4) is indeed the solution to the mean-variance optimization problem with a risk-free asset. For two risky assets with $\boldsymbol{\mu} = (0.08, 0.06)'$, standard deviations $\boldsymbol{\sigma} = (0.18, 0.12)'$, correlation coefficient $\rho_{12} = 0.3$, and a risk-free return of $r_f = 4\%$, determine the equation of the capital market line.
4. Coherent risk measures ρ satisfy the following axioms
   - *Translation invariance*: for all X and $a \in \mathbb{R}$ it holds that $\rho(X + a) = \rho(X) + a$
   - *Monotonicity*: for all X, Y with $X \leq Y$ one has that $\rho(X) \leq \rho(Y)$
   - *Positive homogeneity*: for all X and $a \geq 0$ it holds that $\rho(aX) = a\rho(X)$
   - *Sub-additivity*: for all risks X, Y, we have $\rho(X + Y) \leq \rho(X) + \rho(Y)$

   (a) Motivate these axioms.
   (b) Show that the risk measure $\rho(X) = \mathbb{E}(X) + \alpha \operatorname{Var}(X)$ does not have the monotonicity property.
   (c) Find two random variables X and Y, such that $\mathrm{VaR}(X+Y) > \mathrm{VaR}(X) + \mathrm{VaR}(Y)$, which violates the sub-additivity axiom. (Hint: consider two random variables that can take 0 or one particular positive value.)
5. Compute the confidence level α_1, for which the VaR of a normally distributed random variable corresponds to the ES at some given confidence level α_2.
6. Consider a random pay-out of X, which is 1 EUR or 100 EUR, each with probability $1/2$. What certain minimum payment would you personally ask for in order to prefer the fixed payment over X? Use your answer, if it is smaller than 50.5, to compute the parameter a of an exponential utility function (a will explain your absolute risk aversion), such that the utility function would explain your previous decision. Otherwise justify why there cannot exist a

concave utility function that could be used to explain your answer. For what parameters a would you prefer a fixed amount of 30 EUR over X?

7. Show that minimizing the variance of an investment of given expected return (i.e. (MP3)) corresponds to maximizing the expected utility with a quadratic utility function for arbitrary return distributions.

Exercises with Mathematica

11. Consider two assets with mean returns $\mu_1 = 0.1$ and $\mu_2 = 0.05$, return standard deviations $\sigma_1 = 0.2$ and $\sigma_2 = 0.1$ and correlation coefficient $\rho = 0.25$. You aim to hold a portfolio that satisfies $\sigma_{\text{pf}} = 0.1$. Determine the corresponding optimal Markowitz portfolio. What is the expected return of this portfolio? Determine the portfolio $\boldsymbol{\theta}_{\text{mv}}$ with the smallest variance. What is its expected return? What is the probability of this portfolio to produce a negative return, if you assume returns to be normally distributed?
12. An investor can buy two assets with mean returns μ_1, μ_2 and return variances Σ_{11}, Σ_{22}, and with covariance Σ_{12}. If short-sales (i.e. negative weights) are permitted, show that for any level of initial capital x, mean-variance-optimal portfolios $(\mu_{\text{pf}}^*, \sigma_{\text{pf}}^*)$ form the branch of a hyperbola and identify the efficient frontier. Implement this exercise in Mathematica, and plot the resulting hyperbola for a range of parameter choices. (Hint: the Markowitz problem in this relatively simple market can be turned into an optimization problem in one variable by using the budget constraint, such that a solution can be easily obtained.)
13. Consider a Markowitz model with three possible investments, and

$$\mu_1 = 0.1,\ \mu_2 = 0.05,\ \mu_3 = 0.085,\ \Sigma = \begin{pmatrix} 0.04 & 0.005 & 0.006 \\ 0.005 & 0.01 & 0.0018 \\ 0.006 & 0.0018 & 0.0225 \end{pmatrix}.$$

Use Mathematica to plot the area of the admissible (μ, σ)-combinations and identify the efficient frontier.

14. Consider the model of Exercise 13 and assume that an additional risk-free asset with $r = 0.04$ is available for investment. Identify the capital market line in this example and use Mathematica to plot it together with the original efficient frontier.

Introduction to Credit Risk Models 15

15.1 Introduction

Lending money is one of the core businesses of banks. The income from this business line comes in the form of interest income and we will now discuss why different borrowers will be charged different interest costs in the same lending market. Recall that a loan contract defines a principal amount (or: nominal), a term T by which the principal must be repaid, and a payment schedule according to which payments (interest, principal repayments) of c_i must be made at times t_i ($t_i \leq T, i = 1, ..., n$). The bank is now exposed to the risk of the borrower failing to make payments according to the schedule. This risk is referred to as *credit risk*. The first time τ when the borrower misses a scheduled payment is called *time of default*.[1] If the borrower cannot catch up on missed payments within a *cure period* upon default, the lender can file for the borrower's insolvency with a court. Ideally (for the lender) the borrower will initially have provided collateral (e.g. a mortgage on a house or on land, bank accounts, investment portfolios, valuable paintings, cars, etc.) which the lender can now claim to limit the loss. The *recovery rate* δ gives the fraction of the lender's outstanding claim X_τ at time τ that is ultimately recovered by the lender. Note that the recovery rate will initially be random and only determined during the insolvency process. The payments from the risky (or: *defaultable*) loan/bond can then be summarized as

$$\begin{aligned} &c_i \cdot 1_{\{\tau > t_i\}} \text{ at times } t_i \text{ for } 1 \leq i \leq n, \text{ and} \\ &\delta \cdot X_\tau \cdot 1_{\{\tau \leq T\}} \qquad \text{at time } \tau. \end{aligned}$$

The following sections will discuss several approaches of modeling the distributions of the two random variables τ and δ.

[1]Payment obligations which can lead to a default can also result from other contracts, such as bonds, swaps, options, forwards.

H. Albrecher et al., *Introduction to Quantitative Methods for Financial Markets*, Compact Textbooks in Mathematics, DOI 10.1007/978-3-0348-0519-3_15,

If a bank is approached for a loan, it will assess the credit quality of the potential borrower and offer a loan structure based on the client's risk profile. If this profile indicates higher credit risk, the bank might offer a lower loan (or: principal) amount, more running amortization, or higher interest rates to compensate the risk and it might demand collateral (e.g. a mortgage, a guarantee etc.). A loan is structurally very similar to an insurance contract, where running payments (here: interest) are exchanged for a potential claim amount (here: upon default). The pricing logic for loans is therefore similar to the one for insurance contracts.[2]

For larger borrowings (e.g. to finance business operations or the acquisition of companies or commercial real estate) it can be cheaper to borrow money in the capital markets in the form of bonds (see Chapter 2). In the case of corporate issuers (e.g. Volkswagen, Nestlé, or Microsoft), the bond is called *corporate bond*. Assume a corporate bond is structured as a zero-coupon bond (i.e. no coupons) which promises to pay its nominal N at maturity T. Mostly a bond issue is initially sold in an auction which results in an initial corporate bond price CP. Since the bond is now defaultable, investors will apply a discount rate to the bond cash flows that is higher than the risk-free interest rate r, so that we can write $r + s$, $s > 0$, for the discount rate; s is called *credit spread*. It follows that

$$\mathrm{CP} = e^{-(r+s)T} N, \quad \text{or} \quad s = \frac{-\log(\mathrm{CP}/N)}{T} - r. \tag{15.1}$$

Similarly to the term-structure of the risk-free interest rate $r(t, T)$, the credit spread $s(t, T)$ will also vary for different terms, and the value $\mathrm{CP}(t)$ at time t of a defaultable coupon bond with cash flows c_i at times $t_i > t$ ($i = k, ..., n$, $t_n = T$ and $t_{k-1} \leq t < t_k$) can generally be written as

$$\mathrm{CP}(t) = \sum_{i=k}^{n} e^{-(r(t,t_i)+s(t,t_i))(t_i-t)} c_i. \tag{15.2}$$

15.2 Credit Ratings

In practice credit qualities are grouped in different classes and each class is given a particular grade (or: *credit rating*). Banks will have internal rating systems in place, where they define how to assess and grade the credit quality of a borrower. When issuing a bond, investors will want to understand the credit quality of the potential investment, so that the issuer will ask at least one *rating agency* to assess the credit risk of the issued bond and to publicly communicate the resulting rating to the market. The largest rating agencies include *Moody's*, *Standard & Poor's* (S&P),

[2]Note that apart from the borrower's credit risk, the interest charged will also depend on the costs arising due to regulatory requirements (e.g. capital requirements), the costs at which banks can finance themselves, and the level of competition in the specific lending market.

Quality	Moody's	S&P	Fitch
prime	Aaa	AAA	AAA
high grade	Aa1/Aa2/Aa3	AA+/AA/AA-	AA+/AA/AA-
upper medium	A1/A2/A3	A+/A/A-	A+/A/A-
lower medium	Baa1/Baa2/Baa3	BBB+/BBB/BBB-	BBB+/BBB/BBB-
speculative	Ba1/Ba2/Ba3	BB+/BB/BB-	BB+/BB/BB-
highly speculative	B1/B2/B3	B+/B/B-	B+/B/B-
extremely speculative	Caa1/Caa2/Caa3/Ca	CCC/CC/C	CCC
in default	C	D	DDD/DD/D

Fig. 15.1 (Long-term) rating grade system of Moody's, S&P and Fitch

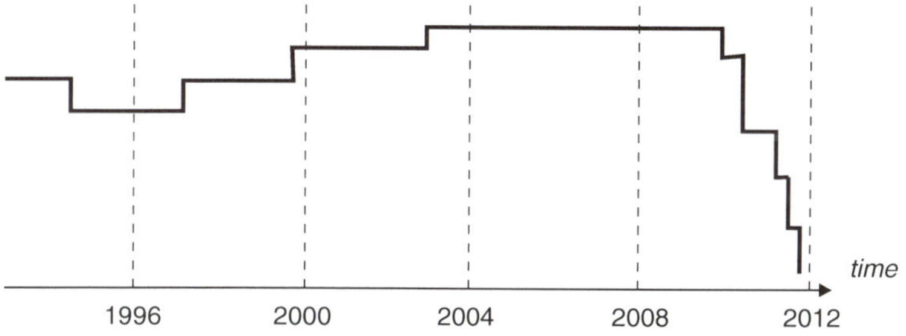

Fig. 15.2 Moody's credit rating history of Greek government debt (1990–2011)

Fitch and DBRS. Each agency applies its own *rating methodology* which it will publish to some extent, to illustrate how particular ratings are attained. In the case of corporates, rating criteria will include qualitative factors (e.g. market position, size, competition, product and geographical diversification, market barriers to entry, distribution system, organizational structure, management) and quantitative factors (e.g. ratios such as earnings/turnover, debt/total assets, debt/equity (also: *leverage*), earnings before interest/interest).

The various rating agencies use similar rating scales (see Figure 15.1). Grades between Aaa/AAA (say: 'triple-A') and Baa3/BBB– are referred to as *investment grades* and indicate good credit quality. Lower grades are called *sub-investment grades* or *speculative*. Ratings as initially published are reviewed regularly and can be *up-* or *downgraded*. A grade step is called *notch*, so that a downgrade by 4 notches from AA– (S&P) would lead to a new rating of BBB+.

Note that the credit rating of an entity can change over time. Figure 15.2 shows the changes in Moody's rating of Greek government debt from 1990 until 2011 (shortly before Greece's debt restructuring in 2012, which was seen as default event by many market participants).[3] This illustrates that even government debt can be far from risk-free, especially if money is borrowed in foreign currency or if the

[3]Source: Moody's (2012), see www.moodys.com.

Table 15.1 One-year global corporate average transition rates (1981–2011) (in %), also referred to as *ratings migration table* (CCC/CC/C is shown as one class, NR indicates that a corporate was not rated any longer at year-end)

From/to	AAA	AA	A	BBB	BB	B	CCC/C	D	NR
AAA	87.19	8.69	0.54	0.05	0.08	0.03	0.05	0.00	3.37
AA	0.56	86.32	8.30	0.54	0.06	0.08	0.02	0.02	4.09
A	0.04	1.91	87.27	5.44	0.38	0.16	0.02	0.08	4.72
BBB	0.01	0.12	3.64	84.87	3.91	0.64	0.15	0.24	6.42
BB	0.02	0.04	0.16	5.24	75.87	7.19	0.75	0.90	9.84
B	0.00	0.04	0.13	0.22	5.57	73.42	4.42	4.48	11.72
CCC/C	0.00	0.00	0.17	0.26	0.78	13.67	43.93	26.82	4.37

government's ability to independently implement monetary or fiscal measures is limited (e.g. when being member of a monetary union).

To put their ratings in a frequency-of-default context, rating agencies track rating changes and defaults, and publish statistics on up-/downgrades. Ratings *migration tables* list e.g. one-year historical transition rates between different rating classes. Table 15.1 shows one-year rating transition rates as reported by S&P (averaged over 1981–2011):[4] over the period 1981–2011, *on average* 8.69% of corporates rated AAA at the beginning of a year had a AA rating by year-end.

It is important to note that ratings migration tables do *not* give forward-looking transition probabilities. For example, S&P reported that for 2011 only 49.02% of initially AAA rates corporates were still AAA by year-end, which is much lower than the long-term average of 87.19% (cf. Table 15.1). Also, S&P publishes sample standard deviations for the transition rates in the long-term average tables. For example, over 1981-2011 this standard deviation was 9.11% for AAA to AAA, and 4.64% for BBB to BBB.

15.3 Structural Models

Structural models interpret defaultable debt as a derivative on the issuer's asset value, and a possible default of the debt is caused *endogenously* (i.e. through reasons lying within the company that cause the asset value to deteriorate). We will now discuss the first and simplest structural model as introduced by R. Merton in 1974.

Let V_t be the value of the assets of a company (including cash, property, patents, inventory etc.) under the physical (i.e. the actual as opposed to the risk-neutral) probability measure. Both the lenders and the owners (equity holders) of the company will claim these assets. In particular, the lenders have a contractual

[4]Source: Standard & Poor's Global Fixed Income Research (2012): 2011 Annual Global Corporate Default Study And Rating Transitions, `www.standardandpoors.com/ratings/articles/`.

senior claim D_t on the company's assets, while equity holders claim the residual asset value after all lenders have been repaid in full. Hence,[5]

$$V_t = D_t + E_t.$$

Merton now models the dynamics of the asset value by a geometric Brownian motion, i.e.

$$dV_t = V_t(\mu dt + \sigma_V dW_t). \tag{15.3}$$

In this simple model, debt is represented by a zero-coupon bond with nominal amount N and maturity T. Since the lender will not be able to recover more than V_T (which is all the company owns) at time T, the value of the debt claim at maturity is

$$D_T = \left\{ \begin{array}{ll} N, & \text{if } V_T \geq N, \\ V_T, & \text{otherwise} \end{array} \right\} = N - (N - V_T)^+. \tag{15.4}$$

The right-hand side of (15.4) now allows one to interpret the pay-off D_T as the contractual (deterministic) nominal amount N, plus the non-positive pay-off of a short position in a European put option on the asset value V_t with strike N.

Using the Black-Scholes formula and applying the put-call parity gives the price at time t of this put option (cf. (7.8)) as

$$\mathrm{CP}_t = e^{-r(T-t)}N - \left(e^{-r(T-t)}N\Phi(-d_{t-}) - V_t\Phi(-d_{t+})\right),$$

with

$$d_{t\pm} = \frac{\log(V_t/N) + (r \pm \frac{1}{2}\sigma^2)(T-t)}{\sigma\sqrt{T-t}}.$$

Formula (15.1) then allows one to calculate the credit spread:

$$s(t,T) = -\frac{\log\left(\Phi(d_{t-}) + \Phi(-d_{t+})e^{r(T-t)}V_t/N\right)}{T-t}. \tag{15.5}$$

Similarly to risk-free interest rates, credit spreads will vary for different terms. The above formula introduces *spread-widening risk* in a natural way. It describes the risk that the value of a corporate bond or loan changes prior to its maturity due to the market spread widening for the borrower (e.g. in case of a worsening of the credit

[5]This model greatly simplifies the balance sheet of the company, since other liabilities (such as reserves or employee pension provisions) are not considered here.

quality). Investors can hence not only suffer losses from bonds if the issuer defaults, but also when selling them in the market before maturity.[6]

The Merton model has proved popular for the modeling of credit risk, and many extensions have been developed which generalize some of its features (e.g. more general liability (debt) structures, different dynamics for V_t, or stochastic interest rates and spreads). Note that the classical model assumes that default only occurs if the assets V_T are insufficient to cover the liability D_T at maturity. In practice, however, banks or bond investors would have the ability to take control early if e.g. the financial situation of the borrower worsens. This is done by agreeing on loan/bond *covenants* which the borrower has to comply with. Such covenants can include qualitative factors (e.g. no change of control of the company, the timely providing of financial reports) or *financial covenants* (e.g. a minimum earnings before interest/interest ratio, a maximum debt/earnings ratio). A *covenant breach* typically results in a *(soft) default* unless the lender *waves* the covenant. A default can then result in anything from temporary operational control of the lender to the restructuring of the loan/bond, to an obligation of the borrower to immediately repay outstanding principal amounts. In 1976 Black and Cox incorporated a financial leverage covenant L_t in a Merton setup by defining

$$L_t = \begin{cases} Ke^{-\gamma(T-t)} & \text{for } t < T, \\ N & \text{for } t = T. \end{cases} \tag{15.6}$$

The time of default is defined as the time when the borrower's asset value V_t drops below L_t for the first time, i.e. $\tau = \inf\{t \leq T : V_t < L_t\}$ (with $\inf \varnothing = \infty$). This is illustrated in Figure 15.3, which shows two possible sample paths of a geometric Brownian motion V_t and the deterministic barrier function L_t. The lower sample path falls below L_t for the first time at $\tau < T$, causing a default, while the upper sample path never crosses L_t and hence survives. In the Black-Cox model, K and γ are chosen such that $L_t \leq Ne^{-r(T-t)}$ (see Exercise 1).

No-arbitrage arguments can be used to compute the credit spread, and the discounted pay-off P to the bond investor as (assuming that a possible liquidation of the company does not trigger any costs):

$$P = e^{-rT}\Big[N1_{\{\tau \geq T\}} - (N - V_T)^+ 1_{\{\tau \geq T\}}\Big] + e^{-r\tau} L_\tau 1_{\{\tau < T\}}.$$

Due to its pay-off structure, the bond can be interpreted as (i) a long-position in a digital knock-out option with a notional of N (the payment at time T if no knock-out event happens) and which pays L_τ at knock-out time $\tau < T$, and (ii) a short-position in a knock-out put option. The knock-out barriers are here given as L_t.

[6]Spread-widening risk caused large losses during and after the 2007/08 credit crisis, when credit spreads significantly increased (or: widened) in fear of future defaults, and bond and loan holders had to take losses when selling credit products in the market.

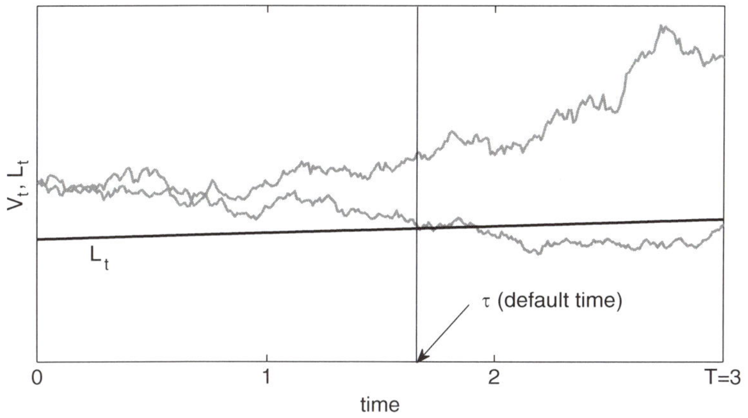

Fig. 15.3 Two sample paths V_t and the barrier L_t in a Black-Cox setup. The lower path defaults, the upper does not

The valuation of barrier options was discussed in Chapter 13 and the same methods can now be applied for the determination of the corporate bond price CP.[7]

Theorem 15.1 (Corporate bond price in the Black-Cox model.). *Let V_t be the value of the assets of a company as in (15.3) and $L_t = e^{-\gamma(T-t)}N$ for $\gamma > r$. Under the assumption $(r-\gamma-\sigma^2/2)^2 > 2(\gamma - r)\sigma^2$, the corporate bond price CP_t at time t is given by*

$$\mathrm{CP}_t = e^{-r(T-t)}N\left(\Phi(d_{t1}) - y_t^{2\alpha}\Phi(d_{t2})\right) + V_t\left(y_t^{1+\alpha+\xi}\Phi(d_{t3}) + y_t^{1+\alpha-\xi}\Phi(d_{t4})\right), \quad (15.7)$$

with

$$y_t = \frac{Ne^{-\gamma(T-t)}}{V_t}, \quad \alpha = \frac{r-\gamma-\sigma^2/2}{\sigma^2}, \quad \xi = \frac{\sqrt{(r-\gamma-\sigma^2/2)^2 - 2(\gamma-r)\sigma^2}}{\sigma^2}$$

and

$$d_{t1} = \frac{\log(V_t/N) + (r-\sigma^2/2)(T-t)}{\sigma\sqrt{T-t}}, \quad d_{t2} = \frac{\log(N/V_t) + (r-2\gamma-\sigma^2/2)(T-t)}{\sigma\sqrt{T-t}},$$

$$d_{t3} = \frac{\log(N/V_t) + (\xi\sigma^2-\gamma)(T-t)}{\sigma\sqrt{T-t}}, \quad d_{t4} = \frac{\log(N/V_t) - (\xi\sigma^2+\gamma)(T-t)}{\sigma\sqrt{T-t}}.$$

[7] We will state here only the case where $K = N$; a more general version of the theorem can be found in Bielecki & Rutkowski [5] or Black & Cox [7].

Remark 15.2. Both the Merton and the Black-Cox models assume that the asset value of the company is a tradable asset. This is one of the main points of criticism for structural models, as for (large listed) companies only the equity (or: stocks) will be priced by the market, but not the company assets.[8] In structural models a company's equity can be interpreted as a call option on the assets. Due to (15.3) and the resulting dependencies, the stock price itself will not follow a Brownian motion.[9] In practice it is a challenge to model complex liability and covenant structures in a structural setup. Structural models are also applied in commercial credit software, such as Moody's KMV package.

15.4 Reduced-Form Models

In structural models a default is triggered by endogenous factors that result in a deterioration in the asset value of the company. In contrast to this, *reduced-form models* (or: *intensity models*) treat default events as being caused by *exogenous* factors. Consider a no-arbitrage market in which the price of a financial product is given by a discounted expectation under the risk-neutral measure $\mathbb{Q}$, thanks to the Fundamental Theorem of Asset Pricing. The cumulative distribution function F_τ under the risk-neutral measure is then given as $F_\tau(t) = \mathbb{Q}[\tau \le t]$. Assume that the probability density function

$$f_\tau(t) = F_\tau'(t) = \lim_{\Delta t \to 0} \frac{\mathbb{Q}[t \le \tau < t + \Delta t]}{\Delta t}$$

exists. The *hazard rate* is then defined as

$$\lambda_\tau(t) := \lim_{\Delta t \to 0} \frac{\mathbb{Q}[t \le \tau < t + \Delta t \mid \tau \ge t]}{\Delta t} = \lim_{\Delta t \to 0} \frac{1}{\Delta t} \frac{F_\tau(t + \Delta t) - F_\tau(t)}{1 - F_\tau(t)} = \frac{f_\tau(t)}{1 - F_\tau(t)}.$$

With $F_\tau(0) = 0$ one finds (see Exercise 4) that

$$F_\tau(t) = 1 - e^{-\int_0^t \lambda_\tau(s)\,ds}. \tag{15.8}$$

The default time τ is then given by

$$\tau = \inf\left\{ t \ge 0 : \int_0^t \lambda_\tau(s)\,ds > \zeta \right\}, \tag{15.9}$$

where ζ is an exponentially distributed random variable with parameter 1 (see Exercise 5). The default time is hence determined exogenously through ζ.

---

[8] The traded price of the equity will however in some cases provide satisfactory information on the market view of the value of the company assets.

[9] A structural model was first applied to the pricing of stock derivatives by Robert Geske in 1979 and leads to the problem of pricing a compound option (i.e. an option on an option).

Let us now consider the calibration of F_τ and $\lambda_\tau(t)$. Since F_τ is the risk-adjusted distribution function, it cannot be estimated from historical data directly, but requires additional information. In Chapter 12 we used prices of traded European options to calibrate stock price models; here we will follow the same procedure, but use corporate bond prices.[10] Consider a corporate bond with nominal amount N and maturity T. For simplicity assume $\delta = 0$. The price CP of the bond is then

$$\mathrm{CP} = \mathbb{E}^{\mathbb{Q}}[e^{-rT} N\, 1_{\{\tau > T\}}] = Ne^{-rT}\, \mathbb{Q}\left[\int_0^t \lambda(s)\, ds \leq \zeta\right] = Ne^{-rT} e^{-\int_0^T \lambda(s)\, ds}. \tag{15.10}$$

Since CP and r can be observed in the market, one finds

$$\int_0^T \lambda(s)\, ds = -\log(\mathrm{CP}/N) - rT.$$

In view of (15.1), λ corresponds to the spread of the bond in case of a constant hazard rate $\lambda(t) \equiv \lambda$.[11]

Since a corporate will often have several bond issues with different maturities outstanding, one will obtain a time-dependent term structure of the default probability (cf. Chapter 1). Formula (15.10) implies that CP corresponds to the price of a risk-free bond with short-rate $r + \lambda(t)$. Note that the price process of a bond with deterministic hazard rate is again deterministic (at least until the time of default), which is of course not the case in practice (as this would neglect the spread risk). As a response to this issue, it is natural to model hazard rates stochastically. Assume that λ_t is a stochastic process that is independent of the random variable ζ. The price CP_t at time t of the corporate bond is

$$\mathrm{CP}_t = Ne^{-r(T-t)} \mathbb{E}^{\mathbb{Q}}\left[e^{-\int_t^T \lambda_s\, ds}\right], \tag{15.11}$$

so that spread-widening risk is now also included in the intensity model. Formula (15.11) shows that CP is determined in an analogous way as a risk-free bond with stochastic short-rate, so that short-rate models can be adapted to the pricing of corporate bonds. In particular, well studied short-rate frameworks can be used for modeling λ_t (cf. Chapter 9). It remains to incorporate the recovery rate upon default in the setup, and it is a common modeling assumption that the recovery is also an exogenous random variable and independent of the time of default.

Remark 15.3. A major point of criticism of intensity models is that they do not reflect all information provided by the stock markets and that they do not consider hedging possibilities (e.g. by taking positions in the company's stock). This has

[10] Other traded credit instruments, as discussed in Section 15.5, can also be applied to calibrate credit risk models.

[11] The parameter $\lambda(t)$ can be interpreted as the current spread rate over an infinitesimal time interval, similar to the current short rate $r(t)$ in interest rate models.

encouraged the development of hybrid models which allow for dependence of λ_t on the stock price S_t. Also note that a realistic description of the recovery rate δ will be crucial yet challenging for the modeling of credit risk.

15.5 Credit Derivatives and Dependent Defaults

Up to this point we have discussed loans and bonds through which the lender or investor is exposed to credit risk. Derivatives on credit risk (more commonly: *credit derivatives*) were developed to allow for the efficient management of credit risk. The most commonly traded structure is the *credit default swap* (CDS), which in principle insures against loss from credit risk of one or more *names* (or: *reference credits*; e.g. Microsoft, the Government of Italy or Nestlé) in exchange for the regular payment of a premium. The CDS market was launched around 1996 and saw a strong growth period over the years 2003 to 2007, with the outstanding (notional) volume reaching around 60 trillion USD by year-end 2007. The 2007/08 credit crisis then brought a major drop in appetite to supply credit insurance, which resulted in a significant decrease in the market size, and the outstanding volume remained in the region of 30 trillion USD in 2009-2011.[12]

Consider the following illustrative example (see Figure 15.4). Investor A holds a corporate bond of company C and would like to limit the risk of taking a loss due to a default of C on its debt. A approaches B, a dealer in credit default swaps, and enters into the following contract: A pays a periodic premium (or: CDS spread) s to B up to time T as long as C has not defaulted on the bond. In turn, B accepts the obligation of paying A a fixed (compensation) amount N in case C defaults on the bond. The CDS contract can specify that the compensation payment is either made at the time of default τ (American style) or at some initially fixed expiry T_{CDS} of the CDS contract. It is common to choose the CDS spread s such that the CDS contract has an initial fair value of 0 (similar to interest-rate swaps). A slightly different structure is given by the *credit default option*, for which the premium is paid in one lump sum when the contract is entered (as opposed to regular CDS premium payments).

CDS contracts can be written on single names or on a portfolio of credits (*multi-name CDS*). Up to this stage we have only considered credit risk models for one bond (or: borrower). However, a bank holds a portfolio of many different credit-risk sensitive contracts, and it is one of its core tasks to model and manage the credit risk arising from this entire portfolio.[13] To assess the credit risk of a pool

[12]Credit derivatives are traded OTC, so that volume estimates are based on figures reported by CDS dealers (mostly banks). Also note that the market volume is normally reported in terms of notional amounts – paid premiums will only account for a fraction of this. The Bank of International Settlement (BIS) collects and publishes OTC derivative volume estimates for the G10 countries and Switzerland in its quarterly reviews, see `www.bis.org`.

[13]The most widely used CDS index in Europe is iTraxx Europe and describes the credit performance of a pool of the 125 most liquid European CDS names. See `www.markit.com`.

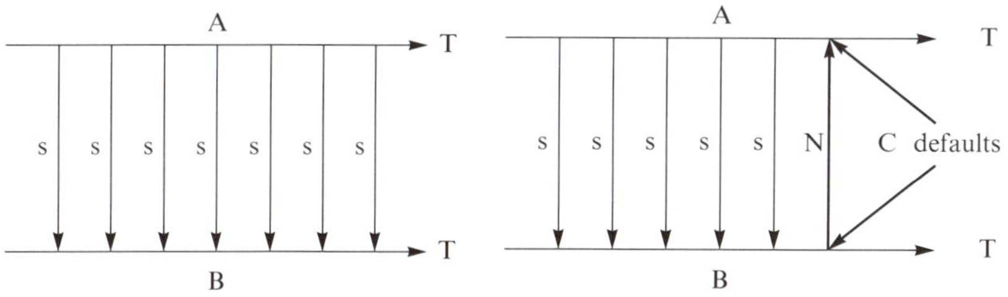

Fig. 15.4 Cash flows between A and B under the CDS contract (notional N, spread s), conditional on C surviving (left) or defaulting (right)

of names, it is now critical to incorporate the default dependence structure of the single names in the model. Similarly, *counterparty risk*, which describes the risk of a counterparty (e.g. the seller of a CDS) in a financial contract to not fulfil its obligations under the contract (e.g. to make the compensation payment upon the default of the reference credit), also requires the modeling of dependent defaults. To incorporate default dependence in structural models, one could start by modeling a multi-variate Brownian motion with positive correlation. This method, however, produces only 'weak' default dependence even for large correlation coefficients, and is therefore inappropriate for incorporating so-called *Armageddon scenarios* (i.e. many defaults occur within a short period of time).

Commonly used credit risk models, such as Moody's *KMV*, JPMorgan's *CreditMetrics*, Credit Suisse's *CreditRisk*$^+$, or intensity models such as the *Duffie-Singleton model* incorporate default dependence in different ways. However, a detailed discussion of the various methods is beyond the scope of this book.[14]

We will now briefly outline another class of financial products that can be seen as derivatives on credit risk: *asset-backed securities (ABS)*. Hereby credit products ('assets') are sold to a company (or: *special purpose vehicle* (SPV)) whose only purpose is holding these assets. To pay for the purchase of the assets, the SPV sells bonds (ABS) to capital market investors who then have a secured claim against the assets owned by the SPV. Practically, this construction transforms many small loans (assets) into larger bonds (ABS), which is also called *re-packaging*. In principle, ABS allow banks to actively manage their credit portfolio by selling certain parts of their loan book, while it allows investors to buy loans (who would otherwise require a banking/lending license in many jurisdictions).

If the asset pool consists of only commercial/residential mortgage-backed loans, ABS are also called *commercial/residential mortgage-backed securities* (short: *CMBS, RMBS*). ABS can generally be issued for pools of any kind of loans, such as auto loans, consumer loans, or credit card receivables. If the pool of assets contains structurally different types of debt products (such as loans, corporate bonds, ABS bonds, mezzanine loans), the ABS are also called *collateralized debt obligations*

[14]Modeling dependence is an active field of research (see references at the end of the chapter).

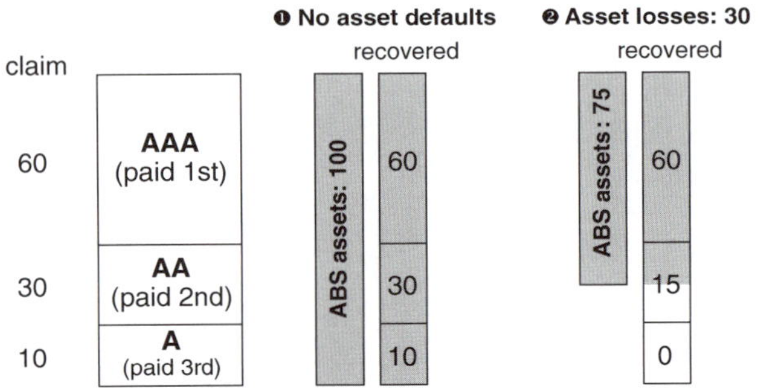

Fig. 15.5 Waterfall of ABS bonds (for the example below): no default on the SPV's assets (left), losses are first covered by the more junior tranches (right)

(CDOs). To attract different bond investor groups, the SPV will typically structure its ABS bond issue into different tranches, i.e. issue low risk senior bonds (paid first; typically rated AAA) as well as higher-risk more junior bonds (paid only when senior bonds have been paid in full; rated lower than the senior tranche, e.g. AA, A, BBB, BB). The defined sequence of coupon and principal payments to the different ABS tranches is called *waterfall* and is defined in the ABS bond contracts. This will be made clear by a short (simplified) example (see Figure 15.5).

Example

Assume an SPV has issued bonds with an aggregate principal amount of 100, composed of AAA (60), AA (30) and A (10) bonds. The waterfall defines that payments are made to AAA before AA before A. If no loans in the SPV's asset pools default, the SPV shall receive an aggregate amount of 100 from the assets, with which it can repay AAA, AA and A bondholders in full (i.e. pay them 60, 30 and 10, respectively). If some of the loans in the SPV's asset pool default, the SPV will take losses on the loan principals in the asset pool, and might only recover a total of 75. When repaying the ABS bondholders, AAA still receives 60 (paid first), AA receives $\min(30, (75 - 60)^+) = 15$ and A receives $\min(10, (75 - 60 - 30)^+) = 0$ through the *sequential* waterfall. AAA only bears losses if the buffer from its *subordinate tranches*, AA and A (i.e. $30+10=40$), is insufficient to cover the SPV's losses. This is referred to as *credit enhancement* of the AAA tranche.

If all ABS bonds are priced at face value, interest rate (credit) spreads will be higher for more junior tranches. AAA tranches can often be structured even for asset pools of only e.g. BBB loans, as one theoretically profits from diversification.

The ABS market had enormously grown in volume up to 2007. However, ABS spreads widened drastically during the 2007/08 financial crisis. This was partly due to a fear of future losses from low-quality assets in ABS pools and of initial underestimation of default dependency within asset pools. ABS values dropped significantly as a result, leading to market value losses for investors holding ABS. Rating agencies revised many ratings of ABS bonds downwards as a reaction to higher expected credit losses, and despite ABS offering a useful structure to allow non-bank investors to participate in the lending market, the future role of the ABS market remains unclear.

15.6 Key Takeaways, References and Exercises

Key Takeaways

After working through this chapter you should understand and be able to explain the following terms and concepts:

- Credit risk: default time τ, recovery rate δ, credit spread s
- Credit ratings: Moody's/S&P/Fitch, AAA—BBB— (investment grade) vs. BB—C (speculative) vs. D (default), ratings migration tables
- Structural models: default caused endogenously, Merton model, financial covenants, Black-Cox extension
- Intensity models: default caused by exogenous events, hazard rate
- Credit derivatives: CDS, credit default spread, counterparty risk
- ABS: ABS vs. CMBS/RMBS vs. CDO, waterfall, credit enhancement, default dependence

References

Merton introduced the valuation of corporate bonds by the use of structural models in his 1974 article [58], and Black & Cox discussed an extension to the model to incorporate safety covenants in [7] in 1976. For a more detailed discussion of structural models and their application to the pricing of stock options, consult Hanke [39]. Bluhm, Overbeck & Wagner [10] and Bielecki & Rutkowski [5] offer a comprehensive overview of the modeling of credit risk.

Exercises

1. In the Black-Cox model, use (a) no-arbitrage arguments and (b) formula (15.7) to show that $s = 0$ if $L_t = Ne^{-r(T-t)}$ for $0 \le t \le T$.
2. Show that the price of a corporate bond the Black-Cox model with $\gamma \to \infty$ converges to the corresponding price in the Merton model.
3. Apply the theorem on the distribution of the first-passage time of a Brownian motion on page 145 to verify (15.7).
4. Prove formula (15.8).
5. Show that the random variable as defined in (15.9) has the cumulative distribution function $F_\tau(t)$.
6. Recall that the buyer of a CDS contract is exposed to the counterparty risk that the seller will not provide the contractual compensation upon the occurrence of a default event of the reference credit. A *credit linked note*, on the other hand, is a bond whose principal is only repaid in full if no credit event occurs until maturity (i.e. it is a pre-funded form of a CDS). Explain how a credit linked note can be constructed from a CDS and a bond.

Exercise with UnRisk

7. Apply the UnRisk command `MakeCreditDefaultSwapCurve`, to describe the default risk of a borrower based on the credit default spread of a CDS. How does the default intensity (hazard rate) change if the recovery rate is varied? (Hint: `HazardRates`)

References

1. M. Aichinger and A. Binder. *A Workout in Computational Finance*. Wiley (to appear), 2013.
2. J. Andreasen, B. Jensen, and R. Poulsen. Eight valuation methods in financial mathematics: The Black-Scholes formula as an example. *Mathematical Scientist*, **23**(1):18–40, 1998.
3. S. Asmussen and P. W. Glynn. *Stochastic Simulation: Algorithms and Analysis*. Springer, New York, 2007.
4. M. Baxter and A. Rennie. *Financial Calculus: An Introduction to Derivative Pricing*. Cambridge University Press, Cambridge, 1996.
5. T. R. Bielecki and M. Rutkowski. *Credit Risk: Modelling, Valuation and Hedging*. Springer Finance. Springer-Verlag, Berlin, 2002.
6. A. Binder and A. Schatz. Finite elements and streamline diffusion for the pricing of structured financial instruments, in: *The Best of Wilmott 2*, P. Wilmott, ed., pages 351–363, Wiley, Chichester, 2006.
7. F. Black and J. Cox. Valuing corporate securities: some effects of bond indenture provisions. *Journal of Finance*, **31**(2):351–367, 1976.
8. F. Black and R. Litterman. Global portfolio optimization. *Financial Analyst Journal*, **48**(5):28–43, 1992.
9. F. Black and M. Scholes. The pricing of options and corporate liabilities. *Journal of Political Economy*, **81**(3):637–654, 1973.
10. C. Bluhm, L. Overbeck, and C. Wagner. *An Introduction to Credit Risk Modeling*. Chapman & Hall/CRC, Boca Raton, FL, 2003.
11. A. Brace, D. Gatarek, and M. Musiela. The market model of interest rate dynamics. *Mathematical Finance*, **7**(2):127–155, 1997.
12. P. Brémaud. *Markov Chains. Gibbs Fields, Monte Carlo Simulation and Queues*. Springer, New York, 2008.
13. D. Brigo and F. Mercurio. *Interest Rate Models—Theory and Practice*. 2nd edition. Springer Finance. Springer-Verlag, Berlin, 2006.
14. M. Capiński and T. Zastawniak. *Mathematics for Finance: An Introduction to Financial Engineering*. 2nd edition., Springer-Verlag, London, 2011.
15. P. Carr and D. B. Madan. Option valuation using the Fast Fourier Transform. *Journal of Computational Finance*, **2**(4):61–73, 1999.
16. V. Chopra and W. Ziemba. The effects of errors in means, variances and covariances on optimal portfolio choice. *Journal of Portfolio Management*, **19**(2):6–11, 1993.
17. D. Christie. Accrued interest and yield calculations and determination of holiday calendars, Technical Report, SWX Swiss Exchange, (2002).
18. R. Cont and P. Tankov. *Financial Modelling with Jump Processes*. Chapman & Hall/CRC, Boca Raton, FL, 2004.
19. R.-A. Dana and M. Jeanblanc. *Financial Markets in Continuous Time*. Springer Finance. Springer-Verlag, Berlin, 2003.
20. F. Delbaen and W. Schachermayer. *The Mathematics of Arbitrage*. Springer Finance. Springer-Verlag, Berlin, 2006.

H. Albrecher et al., *Introduction to Quantitative Methods for Financial Markets*,
Compact Textbooks in Mathematics, DOI 10.1007/978-3-0348-0519-3,

21. E. Derman and I. Kani. Riding on a smile. *RISK*, **7**(2):32–39, 1994.
22. M. Drmota and R. Tichy. *Sequences, Discrepancies and Applications*, *Lecture Notes in Mathematics*, 1651. Springer-Verlag, Berlin, 1997.
23. D. Duffie. *Dynamic Asset Price Theory*. 3rd edition. Princeton University Press, Princeton, NJ, 2001.
24. B. Dupire. Pricing with a smile. *RISK*, **7**(2):18–20, 1994.
25. H. Egger and H. W. Engl. Tikhonov regularization applied to the inverse problem of option pricing: convergence analysis and rates. *Inverse Problems*, **21**(3):1027–1045, 2005.
26. R. J. Elliott and E. P. Kopp. *Mathematics of Financial Markets*. 2nd edition. Springer, New York, 2005.
27. H. W. Engl. *Integralgleichungen*. Springer-Verlag, Vienna, 1997.
28. H. W. Engl, M. Hanke, and A. Neubauer. *Regularization of Inverse Problems*. Kluwer, Dordrecht, 1996.
29. H. W. Engl. Calibration problems—an inverse problems view. *WILMOTT Magazine*, pages 16–20, 2007.
30. F. Fabozzi, P. N. Kolm, D. Pachamanova, and S. M. Focardi. *Robust Portfolio Optimization and Management*. Wiley, Hoboken, NJ, 2007.
31. W. Feller. *An Introduction to Probability Theory and Its Applications*, volume 2. Wiley, New York, 1970.
32. E. Fernholz. *Stochastic Portfolio Theory*. Springer, New York, 2002.
33. D. Filipović. *Term-Structure Models: A Graduate Course*. Springer-Verlag, Berlin, 2009.
34. H. Föllmer and A. Schied. *Stochastic Finance: an Introduction in Discrete Time*. 3rd edition. Walter de Gruyter & Co., Berlin, 2011.
35. J.-P. Fouque, G. Papanicolaou, and K. R. Sircar. *Derivatives in Financial Markets with Stochastic Volatility*. Cambridge University Press, Cambridge, 2000.
36. G. Fusai and A. Roncoroni. *Implementing Models in Quantitative Finance: Methods and Cases*. Springer Finance. Springer-Verlag, Berlin, 2008.
37. M. B. Garman and S. W. Kohlhagen. Foreign currency option values. *Journal of International Money and Finance*, **2**(3):231–237, 1983.
38. P. Glasserman. *Monte Carlo Methods in Financial Engineering*. Springer, New York, 2004.
39. M. Hanke. *Credit Risk, Capital Structure, and the Pricing of Equity Options*. Springer-Verlag, Wien, 2003.
40. S. L. Heston. A closed-form solution for options with stochastic volatility with applications to bond and currency options. *Review of Financial Studies*, **6**(2):327–343, 1993.
41. J. C. Hull. *Options, Futures, and other Derivatives*. 5th edition. Prentice Hall, Upper Saddle River, NJ, 2002.
42. J. C. Hull and A. D. White. Numerical procedures for implementing term structure models II: two-factor models. *Journal of Derivatives*, **2**(2):37–48, 1994.
43. J. C. Hull and A. D. White. Using Hull-White interest rate trees. *Journal of Derivatives*, **3**(3):26–36, 1996.
44. B. Kaltenbacher, A. Neubauer, and O. Scherzer. *Iterative Regularization Methods for Nonlinear Ill-Posed Problems*. Radon Series on Computational and Applied Mathematics, de Gruyter, Berlin, 2008.
45. I. Karatzas and S. E. Shreve. *Brownian Motion and Stochastic Calculus*. 2nd edition. Springer-Verlag, Berlin, 1998.
46. I. Karatzas and S. E. Shreve. *Methods of Mathematical Finance*. Springer, New York, 1998.
47. R. Korn and E. Korn. *Optionsbewertung und Portfolio-Optimierung*. Vieweg, Braunschweig, 1999.
48. R. Korn, E. Korn, and G. Kroisandt. *Monte Carlo Methods and Models in Finance and Insurance*. Chapman & Hall/CRC, Boca Raton, FL, 2010.
49. D. Lamberton and B. Lapeyre. *Introduction to Stochastic Calculus Applied to Finance*. 2nd edition. Chapman & Hall/CRC, Boca Raton, FL, 2008.
50. S. Larsson, V. Thomée. *Partial Differential Equations with Numerical Methods*, Springer, Berlin, 2003

51. R. Lee. Option pricing by transform methods: extensions, unification and error control. *Journal of Computational Finance*, **7**(3):51–86, 2004.
52. A. L. Lewis. *Option Valuation under Stochastic Volatility*. Finance Press, Newport Beach, CA, 2000.
53. F. A. Longstaff and E. A. Schwartz. Valuing American options by simulation: a simple least-squares approach. *Review of Financial Studies*, **14**(1):113–147, 2001.
54. R. Lord and C. Kahl. Optimal Fourier inversion in semi-analytical option pricing. *Journal of Computational Finance*, **10**(4):1–30, 2007.
55. D. Luenberger. *Investment Science*. Oxford University Press, New York, 1997.
56. A. J. McNeil, R. Frey, and P. Embrechts. *Quantitative Risk Management: Concepts, Techniques, and Tools*. Princeton University Press, Princeton, NJ, 2005.
57. R. Merton. Theory of rational option pricing. *Bell Journal of Economics and Management Science*, **4**(1):141–183, 1973.
58. R. Merton. The pricing of corporate debt: the risk structure of interest rates. *Journal of Finance*, **29**(2):449–470, 1974.
59. H. Niederreiter. *Random Number Generation and Quasi-Monte Carlo Methods*. Society for Industrial and Applied Mathematics (SIAM), Philadelphia, PA, 1992.
60. B. Øksendal. *Stochastic Differential Equations: An Introduction with Applications*. 6th edition. Springer-Verlag, Berlin, 2010.
61. A. Pascucci and W. J. Runggaldier. *Financial Mathematics: Theory and Problems for Multi-Period Models*. Springer-Verlag, Italia, Milan, 2012.
62. G. Pflug and W. Römisch. *Modelling, Measuring and Managing Risk*. World Scientific, Hackensack, NJ, 2007.
63. E. Platen and D. Heath. *A Benchmark Approach to Quantitative Finance*. Springer Finance. Springer-Verlag, Berlin, 2006.
64. R. Rebonato. *Modern Pricing of Interest-Rate Derivatives: The LIBOR Market Model and Beyond*. Princeton University Press, Princeton, NJ, 2002.
65. L. C. G. Rogers and D. Williams. *Diffusions, Markov Processes, and Martingales. Vol. 1*. Cambridge University Press, Cambridge, 1994.
66. H.-G. Roos, M. Stynes, and L. Tobiska. *Numerical Mehods for Singularly Perturbed Differential Equations. Convection-Diffusion and Flow Problems*. Springer-Verlag, Berlin, 1996.
67. W. Schachermayer and J. Teichmann. Wie K. Itô den stochastischen Kalkül revolutionierte. *Internationale Mathematische Nachrichten*, Wien, **205**:11–22, 2007.
68. B. Scherer and R. D. Martin. *Introduction to Modern Portfolio Optimization*. Springer, New York, 2005.
69. W. Schoutens. *Lévy Processes in Finance*. Wiley, New York, 2003.
70. R. U. Seydel. *Tools for Computational Finance*. 4th edition. Springer-Verlag, Berlin, 2009.
71. S. E. Shreve. *Stochastic Calculus for Finance I: The Binomial Asset Pricing Model*. Springer-Verlag, New York, Berlin, 2005.
72. S. E. Shreve. *Stochastic Calculus for Finance II: Continuous-Time Models*. Springer-Verlag, Berlin, 2004.
73. J. Topper. *Financial Engineering with Finite Elements*. Wiley, Chichester, 2005.
74. Triennial Central Bank Survey. Bank for International Settlements, 2010.
75. P. Wilmott. Introduces *Quantitative Finance*. Wiley, Chichester, 2007.
76. W. Zulehner. *Numerische Mathematik: eine Einführung anhand von Differentialgleichungsproblemen. Band 1. Stationäre Probleme*. Mathematik Kompakt. Birkhäuser Verlag, Basel, 2008.

Index

H. Albrecher et al., *Introduction to Quantitative Methods for Financial Markets*, Compact Textbooks in Mathematics, DOI 10.1007/978-3-0348-0519-3,

Printed by Printforce, the Netherlands